Strategic Thinking in Bets

Decision Quality for Investors and Traders

Joseph Trek

Published by NobleTrex Press

For permissions and other inquiries, write to:

P.O. Box 3132, Framingham, MA 01701, USA

Contents

Introduction

Imagine you are at a party. It's late, the music is loud, and you are trapped in a corner listening to a neighbor-let's call him Steve-brag about his portfolio. Steve tells you, with the unearned confidence of a man who has had three martinis, that he put his entire retirement savings into a cryptocurrency named after a dog, or perhaps a struggling video game retailer, because he "had a feeling."

He pulls out his phone. He shows you the chart. He is up 400 percent in two weeks. He is beaming. He looks at you, the diligent investor who diversifies, reads balance sheets, and understands what a standard deviation is, and he offers a pitiful smile. "You play it too safe," he says.

In that moment, your brain screams two contradictory things. First: *Steve is an idiot.* Second: *Steve is winning, and I am losing.*

This book is about resolving that tension. It is about understanding why Steve's bank account is flush while his decision-making process is bankrupt, and why mimicking him is the fastest way to destroy your financial future. But more importantly, it is about the most difficult, counterintuitive, and profitable skill you can learn in the financial markets: the ability to decouple your decisions from your results.

We are hardwired to believe that good outcomes come from good decisions and bad outcomes come from bad decisions. In most of life, this is true. If you touch a hot stove and burn your hand, the pain is a reliable signal that touching the stove was a bad idea. If you practice the piano every day and learn a concerto, the applause is a reliable signal that your practice was effective.

But the financial market is not a piano, and it is certainly not a stove. It is a probabilistic field of fog where you can make a terrible decision-like leveraging your life savings on a hunch-and get rich, or make a mathematically perfect decision and lose money.

This creates a dangerous feedback loop. The market frequently rewards recklessness and punishes prudence, at least in the short term. If you judge your strategy solely by the number at the bottom of your brokerage statement at the end of the month, you are letting randomness teach you how to trade. You are effectively the drunk driver who made it home safely and concluded that driving drunk is a skill.

To survive and thrive as an investor or trader, you must abandon the search for certainty and embrace the art of the bet.

The Illusion of Certainty

Most financial literature sells you a map. It promises that if you follow X indicator, or buy Y asset class, or use Z algorithm, you will arrive at the treasure. These books treat the market like a physics problem-a solvable equation where input A always leads to output B.

But the market is not physics. It is a complex adaptive system driven by human psychology, infinite variables, and incomplete information. It is less like chess-where all the pieces are visible and the rules are fixed-and more like poker, where you can't see your opponent's cards, the deck is constantly being reshuffled, and sometimes the casino changes the rules mid-hand.

When you stop viewing investing as a search for the "right" answer and start viewing it as a series of bets, everything changes. You stop asking, "Will this stock go up?"-a question that demands a yes/no certainty that doesn't exist. Instead, you start asking, "What is the probability of this trade working out? What is the payout if I'm right? How much do I lose if I'm wrong? And is the risk worth the reward?"

This shift from *prediction* to *probability* is not semantic; it is structural. It requires you to rebuild your mind.

You have to get comfortable with the idea that you can be "right" about a trade-meaning you identified a situation with a 70% probability of success-and still lose money. A 70% chance of rain still leaves a 30% chance of sunshine. If it doesn't rain, the meteorologist wasn't wrong; the probability just manifested the minority outcome. In the markets, however, we fire the meteorologist. We abandon strategies that go through a drawdown, even if the logic remains sound. We chase the strategies that are hot, even if their success is statistically unsustainable.

We are going to fix that.

The Architecture of a Good Bet

A strategic bet is not a gamble. A gamble is throwing money at an outcome where you have no edge, hoping for luck. A strategic bet is a calculated allocation of capital where the expected value is positive, the downside is defined, and the sizing ensures you can survive long enough to realize your edge.

In the chapters ahead, we are going to dismantle the way you currently look at your portfolio. We aren't going on a voyage or a journey; we are going into the engine room of decision quality.

We begin by examining the foundation of your thinking. You'll learn to distinguish between *Process* and *Outcome*. This is the mental wedge that separates professionals from amateurs. Professionals obsess over the quality of their thesis and the discipline of their execution; amateurs obsess over the profit and loss (P&L) of the last five minutes. You will learn to recognize when "noise" is masquerading as a signal, and how the chaotic nature of market variance can fool you into thinking you are a genius or an idiot, when you might be neither.

From there, we have to talk about the math. Don't worry-this isn't a textbook on stochastic calculus. But you cannot be a serious capital allocator without speaking the language of probability. We will explore how to assign actual percentages to your beliefs. It is easy to say, "I'm bullish on gold." It is much harder, and more useful, to say, "I believe there is a 65% probability gold outperforms the S&P 500 over the next six months." The moment you quantify your belief, you create something you can test, measure, and improve. We will also dive into Bayesian updating-the scientific method of changing your mind. When the facts change, most people dig in their heels. You will learn to update your priors without bruising your ego.

Of course, even the best mathematical models reside in biological hardware that hasn't had a firmware update in 50,000 years. We are walking bundles of cognitive biases, prone to overconfidence, hindsight bias, and the narrative fallacy. We love stories more than statistics. We see patterns in random clouds. We will spend time navigating the "Human Element," not just to identify these biases, but to build guardrails against them. You cannot willpower your way out of a cognitive bias any more than you can willpower your way out of color blindness, but you can build systems that correct for your flawed vision.

The Edge and the execution

Once we have the mindset, we need the methodology. How do you actually find an edge? In an era where algorithms trade in microseconds and hedge funds spend billions on alternative data, can you still compete? The answer is yes, but not by playing their game. You don't beat the house by being faster; you beat them by being more robust. We will look at how to construct a hypothesis, how to test it without lying to yourself with "backtesting porn," and how to differentiate between a lucky streak and a genuine market advantage.

But knowing what to buy is only ten percent of the battle. The other ninety percent is how much to buy.

This is where the concept of "Managing the Bet" comes in. I will show you that position sizing is more important than stock picking. You can be the world's greatest analyst, but if you bet too big, a standard run of bad luck will bankrupt you. Conversely, you can be a mediocre analyst with excellent risk management and compound money for decades. We will explore the delicate balance of the Kelly criterion, the mechanics of diversification (real diversification, not just owning ten different things that all crash together), and the absolute necessity of defining your "uncle point"-the price at which you admit you are wrong.

Finally, we will close the loop. A bet isn't finished when you close the trade. That is simply the end of the financial transaction; the intellectual transaction is just beginning. We will build a framework for feedback-journaling, post-mortems, and performance attribution-that allows you to strip away the luck and see the skill (or lack thereof) beneath. This is how you stop making the same mistakes for ten years and calling it "experience."

The Price of Admission

Before we turn the page, I need to be honest about the cost of this approach.

Thinking in bets is uncomfortable. It requires you to abandon the comfort of "knowing" what will happen next. It forces you to say "I don't know" more often. It demands that you take responsibility for your process, even when the outcome is painful, and that you remain humble even when the outcome is glorious.

It also alienates you from the crowd. When your friends are panicking because the market dropped 5%, you will be looking at your pre-calculated volatility bands and seeing a non-event. When the media is hyperventilating about the "next big thing," you will be dryly calculating the base rate probability of success. You will be the person at the cocktail party who doesn't have a hot stock tip, but who also doesn't blow up their account every seven years.

There is a freedom in this uncertainty. When you stop trying to predict the future and start preparing for the distribution of outcomes, the anxiety of the market dissipates. You stop worrying about being wrong, because "wrong" is just part of the probability set you already accounted for. You stop envying Steve and his dog-money, because you know that without a rigorous process, the market will eventually reclaim those winnings with interest.

The market is a ruthless teacher, but it is fair in the long run. It transfers wealth from those who seek certainty to those who manage uncertainty. It transfers wealth from the gambler to the house.

This book is the blueprint for building your own house. Let's begin.

Chapter 1

The Foundations of Thinking in Bets

Imagine going all-in with a pair of twos against a pair of aces in poker. If you get lucky and win the hand, does that make you a genius? In the financial markets, we often mistake dumb luck for skill and bad breaks for incompetence. This chapter strips away the illusion of certainty to reveal the mechanics of decision-making. We will learn to separate the quality of our process from the immediate results, define what 'winning' actually looks like for your specific circumstances, and confront the mathematical realities-like volatility and friction-that govern how money truly grows.

1.1 Process Over Payoffs: Redefining Decision Quality

A trade closes green and your brain throws confetti. A trade closes red and your brain holds a trial. Same you, same market, same setup—yet the verdict feels prewritten by the P&L. That reflex is so natural it masquerades as wisdom: *winning means you were right; losing means you were wrong*. Markets adore this confusion because they pay out on a schedule that has only a casual relationship with your competence.

The trap has a name: *resulting*. Resulting is the habit of judging the quality of a decision by the quality of its outcome. It's a seductively tidy shortcut. It turns a messy world of partial information, shifting probabilities, and hidden variables into a simple scoreboard. But it's also how investors train themselves to repeat bad bets and abandon good ones. If you only reward yourself when you get paid, you'll start optimizing for emotional relief—*not* for expected value.

Resulting thrives in domains where the feedback is loud and the truth is quiet. Trading is exactly that kind of domain. The market gives you immediate, numeric feedback, which feels objective—down to the cent, down to the millisecond. Meanwhile, the thing you actually need to evaluate-the logic of the trade *given what was knowable at the time*-is invisible unless you capture it deliberately. If you don't, your future self will "remember" a story that flatters the outcome. A winner becomes a "great read." A loser becomes a "stupid entry." The story changes, but the underlying information set you had in the moment does not.

There's a darker implication here: resulting doesn't just grade your performance; it shapes your identity. A string of lucky wins makes you feel gifted. A string of unlucky losses makes you feel broken. Both feelings are dangerous. Confidence built on luck invites leverage. Shame built on variance invites paralysis. Markets don't merely test your strategy—they test your ability to keep your strategy separate from your mood.

The cleanest way out is to split the world into two buckets: *process* and *outcome*. Process is what you can control: your research, your entry criteria, your sizing, your risk limits, your patience, your ability to say "I don't know." Outcome is what the world delivers: price paths, headlines, flows, correlations that snap, liquidity that vanishes at the worst possible moment. Confusing those buckets is like grading a pilot by whether the plane hit turbulence. Turbulence matters, sure—but it is not a verdict.

Once you accept that outcome and decision quality can diverge, you unlock a more accurate map of your results. The same outcome can come from radically different decision processes, and the same process can produce radically different outcomes. That's not philosophy; it's arithmetic. In a probabilistic environment, you can do everything "right" and still lose. You can do everything "wrong" and still win. If you don't make room for those facts, you'll hallucinate skill where there is none—and fail to recognize skill when it's present.

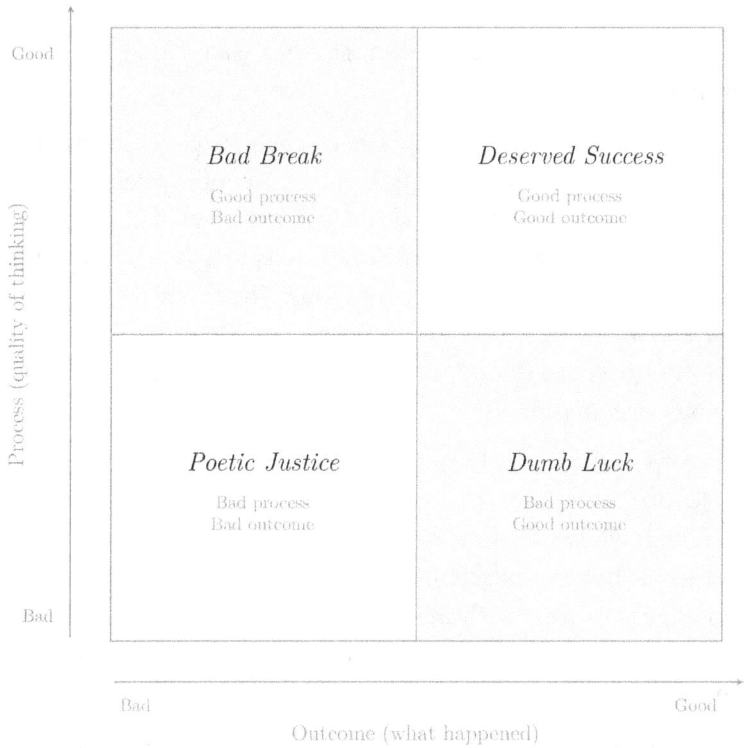

A matrix like this sounds almost too simple-until you try to live inside it. Most people, most of the time, collapse the top row into "good" and the bottom row into "bad," as if outcomes are a perfect lie detector for decision-making. They aren't. Markets don't certify your process; they merely sample it. One trade is a single data point from a distribution. Treating it as a verdict is like tasting one grain of salt and declaring the soup ruined.

The two quadrants that cause the most long-term damage are *Dumb Luck* and *Bad Break,* because they teach the wrong lessons. Dumb

Luck is the worst teacher in finance. It hands you money for a bad reason, and money is persuasive. It makes the risky feel "validated." It makes sloppiness feel "intuitive." A trader buys a biotech on a rumor with no sizing discipline, the drug trial pops, and suddenly the brain writes a new rule: *rumors are alpha*. The real rule is harsher: *you survived a decision that should have hurt you*. Dumb Luck turns survivorship into a strategy.

Bad Break is more insidious because it attacks your ability to trust yourself. You followed your rules, you sized appropriately, you had a thesis with a time horizon, and then an unexpected macro headline detonates the position. Bad Break whispers, *your rules don't work*. That's how people become superstitious. They start trading the last outcome instead of the next opportunity. They get smaller after losses not because the opportunity set changed, but because their nervous system did.

Notice what's missing from this framework: moral judgment. "Poetic Justice" is not the market punishing you; it's the market being indifferent while your process carried hidden explosives. "Deserved Success" is not the market rewarding you; it's the market allowing your edge to express itself this time. The market is not a teacher with a curriculum. If you want to learn, you have to build the curriculum yourself.

That curriculum begins with a question that feels almost unfairly practical: *What did you believe, and what did you know, at the time you acted?* Not what you *feel* now. Not what seems obvious in hindsight. At the time. Decision quality lives in the moment before the click—the last moment where you had agency. After that, you are mostly a passenger.

The easiest way to see the difference is to borrow an example from poker, because poker is finance with the emotions turned up and the math kept honest. Suppose you get the money in as a 70% favorite. You lose the hand. Was the decision bad? Of course not. You made a positive expected value bet. Over time, that's how you win.

If you punish yourself for that loss, you'll gradually train yourself to avoid the very situations that create profitability. You will become "disciplined" in the way that ruins you: disciplined about avoiding discomfort.

Now flip it. Suppose you get the money in as a 30% underdog because you were impatient, bored, and chasing. You win the hand. Was the decision good? No. You got paid *this time*, but you bought a losing lottery ticket. If you celebrate that win without qualification, you'll repeat the behavior until the variance that blessed you becomes the variance that breaks you. The casino loves when you confuse those two.

Trading isn't poker in one crucial way: in poker, you can often estimate the odds. In markets, the odds are fuzzy. Your "70% favorite" is more like "I think I'm slightly favored, unless I'm missing something important." That fuzziness is not a reason to give up on process; it's the reason process matters. When probabilities are uncertain, your protection comes from *repeatability*: having a method that prevents you from letting today's mood become tomorrow's portfolio.

A good process doesn't require you to predict correctly. It requires you to *size your ignorance*. That phrase sounds poetic, but it's literal. If you're uncertain, you size smaller. If the trade depends on a catalyst you can't model, you define the loss you're willing to pay for that optionality. If the setup is crowded, you respect the fact that exits can become one-way doors. Process is risk management wearing the costume of humility.

One reason investors resist process-based thinking is that it feels like surrender. People want the market to be a meritocracy that pays on performance. The uncomfortable truth is that markets pay on outcomes, and outcomes are a blend of skill and chance. Skill is real—but it expresses itself statistically, not ceremonially. If you need every trade to confirm your competence, you'll be at war with randomness, and randomness has infinite ammunition.

This is where the idea of "edge" gets abused. Edge is not a feeling of confidence. Edge is not the adrenaline of a fast tape. Edge is a small, durable advantage—one that might not show up for weeks, months, or even years, depending on your horizon. A real edge often feels boring in the moment because it is systematic. It looks like: "I do this one thing slightly better than average, and I do it again, and again, and again." Resulting is the habit that kills systematic behavior because it demands drama. It demands immediate proof.

So what does "good process" look like in practice, in language that survives contact with an actual trading day? Start with the simplest description: *good process means you can explain why the bet made sense before you knew the ending*. That explanation doesn't need to be ornate. It does need to be anchored to something more stable than vibes. A thesis. A valuation gap. A structural flow. A risk premium. A mean reversion regime. A catalyst with a defined window. Any of these can be legitimate-if you can articulate the logic and map it to a plan for being wrong.

Being wrong is the part most people avoid writing down. They prefer a thesis that reads like a forecast: "This will go up." A process reads like a conditional: "If *this* is true, then price is likely to do *that*; if price does *this* instead, my thesis is probably wrong, and I will do *that*." The difference is that a forecast flatters you, while a conditional protects you.

A useful way to pressure-test whether you're evaluating process rather than outcome is to run a mental "alternate universe" audit. Imagine the same trade, same entry, same stop, same sizing—only the outcome is flipped by an exogenous coin toss. Would you still approve of the decision? If your answer depends on the P&L, you're not evaluating decision quality; you're evaluating your emotional reaction to money.

The mind protests: *but money is the point*. Yes. Money is the score. It is not the coaching. In performance sports, you don't redesign your training plan after every game based on whether you won by three points or lost by three points. You study film. You track conditioning. You work on what you can repeat. In markets, "film" is your pre-trade reasoning, your sizing logic, and your execution notes. If you don't record them, you are asking memory—an instrument optimized for storytelling—to serve as your compliance department.

That's why professionals who last tend to keep some form of decision journal, even if it's scrappy. Not a diary of feelings. A record of claims. What you believed. What would change your mind. What you expected to happen and over what time frame. What risk you were paying (stop distance, option premium, drawdown tolerance). What you thought the distribution looked like-fat left tail or capped loss, convex or linear. When you write those things down, you make it possible to learn from outcomes without being hypnotized by them.

Learning is not "I lost, therefore it was bad." Learning is closer to: "Given the information I had, was my thesis coherent? Were my assumptions reasonable? Did I size for the uncertainty? Did I manage the position according to plan? Did I get new information that should have updated my probabilities?" That last word—*probabilities*—makes people uneasy because it feels too mathematical for a messy world. But you already use probabilities; you just do it unconsciously. You already act as if some outcomes are more likely than others. Process is making that implicit calculus explicit enough to inspect.

Once you start thinking this way, a surprising thing happens: losses become less toxic. They still sting-money is real-but they stop automatically implying personal incompetence. And wins become less intoxicating. They still feel good, but they stop automatically implying genius. That emotional moderation isn't a personality trait; it's a byproduct of better accounting. You stop depositing all your self-worth into a bank that settles in real time.

The biggest payoff of process thinking is not serenity. It's *signal extraction*. You're trying to detect a faint edge inside a hurricane of variance. Resulting is like trying to measure temperature with a thermometer that changes its reading based on whether you like the weather. A process lens makes you ask: "What was the expected value of the decision, given what was knowable?" That expected value might be positive even when you lose, and negative even when you win. If you can't say that sentence with a straight face, you're not running a strategy—you're running a mood.

This also clarifies why "being right" is a childish goal in markets. Being right is binary; markets are not. You can be directionally right and still lose because you were early, because you sized too large, because you chose an instrument with hidden path dependence, because you underestimated how far wrong it could go before it went right. You can be directionally wrong and still make money because you had convexity, because you traded relative value, because you managed the position dynamically. Right/wrong is a poor taxonomy. Process/outcome is a better one.

Take a simple scenario: you buy a stock because you think it's undervalued, but you ignore the fact that you need the cash in six weeks. The stock drops 15% in a broad risk-off move. You sell at the bottom because you must. The outcome is a loss, but the deeper error isn't "picked the wrong stock." The deeper error is *mismatched bet to constraints*. That's a process failure that can hide inside a winning trade too. If the stock had rallied, you might have learned the wrong lesson: "I'm good at picking undervalued stocks," while the real lesson was: "I got away with taking liquidity risk I couldn't afford."

Or consider the opposite: you short a euphoric name with a strong valuation argument but no plan for borrow costs, squeeze risk, or the simple fact that irrational markets can stay irrational longer than your margin. You get stopped out on a face-ripping squeeze. You were "right" in some conceptual sense, but your process was incomplete. The market didn't invalidate your thesis; it exposed your bet structure. Process thinking forces you to treat implementation as part of the decision, not an administrative detail.

There's an uncomfortable corollary: sometimes the most valuable "wins" are the ones you refuse to celebrate. You make money, but you mark it as Dumb Luck because your entry was impulsive or your sizing was reckless. You treat the profit as a warning label, not a trophy. That is emotionally countercultural. It also might be the most investor-like thing you can do: separating dopamine from feedback.

Similarly, some of the most valuable "losses" are the ones you refuse to regret. You lose money, but you mark it as a Bad Break because the thesis was sound and the risk was appropriately priced. You don't romanticize the loss; you simply refuse to let it bully you into rewriting a good playbook. That is how you avoid turning normal variance into strategy drift.

None of this means outcomes don't matter. Outcomes matter in the only way that matters: as data. But data must be interpreted through a model. If your model is "win = good decision," you will learn nonsense. If your model is "process quality and outcome are related but not identical," you can start doing something adults do: updating beliefs. Over enough repetitions, outcomes help you test whether your process produces more Deserved Success than the baseline would predict. But you need enough repetitions, and you need consistent tagging of which quadrant you were actually in.

If you want a single sentence to carry into your next trade, make it this: *I am not trying to be right; I am trying to make good bets*. Good bets are the kind you would take again in an alternate universe where the coin flip lands differently. Good bets are the kind that, over a large sample, don't require you to be a hero-only to be consistent. Good bets are built to survive your own psychology, because the market will happily exploit any part of you that confuses a paycheck with proof.

Process over payoffs isn't a moral stance. It's a competitive advantage. Most participants are trained—by apps, by social media, by the visceral clarity of profit and loss—to chase the emotional simplicity of the scoreboard. The investor who learns to grade decisions in the quiet space *before* the outcome is known is doing something rarer: building a system that can actually improve.

And improvement is the whole game. Not today's P&L. Not a single heroic trade. The slow, almost unglamorous act of turning experience into judgment—without letting randomness take credit or take blame.

1.2 Setting the Stage: Your Goals and Constraints

A "good bet" is not a universal object, like gravity. It's more like a shoe: it only makes sense on a particular foot, for a particular walk, on a particular day. Put differently, a trade can be brilliantly analyzed and still be a terrible decision for *you*. Not because the market disagrees, but because your life does.

This is where a lot of investing advice quietly commits fraud. It talks as if the only thing that matters is whether an asset has attractive expected returns. But investors don't live inside expected returns. They live inside calendars, paychecks, tax bills, mortgages, tuition schedules, and the very real emotional cost of watching a portfolio swing around. Decision quality is always conditional on constraints. Ignore that and you'll end up with the financial equivalent of a race car with no brakes: impressive right up until the first corner.

The fastest way to understand constraints is to stop thinking of them as limitations and start thinking of them as *terms of survival*. You can have the best "edge" in the world and still blow yourself up by violating your own boundary conditions. That's not bad luck; it's self-sabotage with a spreadsheet. The market is allowed to be volatile. You are not allowed to be fragile.

Most investors can name their goals in a vague, socially acceptable way: "retire comfortably," "build wealth," "make my money work." The vagueness is costly. Vague goals invite vague bets, and vague bets are hard to evaluate. "Comfortably" doesn't tell you what drawdown you can tolerate without panic-selling. "Build wealth" doesn't tell you whether a 40% equity allocation is too timid or a 90% allocation is reckless. If you want decisions you can actually grade, goals need a spine: a time horizon, a dollar target (even a rough one), and an honest description of what failure looks like.

Time horizon is the first spine vertebra. Not "long-term" as a personality trait, but specific horizons tied to real obligations. There's a world of difference between money you might need in 18 months and money you won't touch for 18 years. In markets, time is shock absorber. The longer your horizon, the more you can let volatility be noise instead of an emergency. The shorter your horizon, the more volatility becomes a *liquidity event*—a forced sale disguised as "risk."

That idea—forced selling—doesn't get enough respect because it sounds like a niche problem. It isn't. Forced selling is how good strategies die in ordinary households. You can be correct about the long-run merits of owning risky assets and still get crushed if you have to liquidate during a drawdown to pay for something non-negotiable. The market doesn't care that your thesis was "right over five years" if your bank account cares that rent is due on Friday.

Liquidity needs are the second vertebra. Liquidity is simply the ability to turn investments into spendable cash *when you choose*, not only when the market feels generous. It's the difference between owning a financial asset and being owned by it. Investors tend to treat liquidity like an optional optimization: "I'll keep a little cash drag, I guess." In reality, liquidity is emotional insurance with a very practical payoff: it reduces the odds that you'll be a seller at the worst possible time.

A boring emergency buffer-money that is not trying to be clever-often does more for long-term returns than a clever stock pick. That sounds like an insult to finance, but it's just math plus psychology. The buffer keeps your high-volatility assets from being repurposed into a checking account during a crisis. It protects your investing portfolio from your life. And your life will eventually throw something at you: job loss, medical bills, a move, a family obligation, or simply a year where expenses arrive in clusters like bad news.

Risk tolerance is the third vertebra, and it's commonly misunderstood because people treat it as a mood. "I'm aggressive." "I'm conservative." Those are vibes, not constraints. A more useful definition separates two things that often get tangled: *willingness* to take risk and *ability* to take risk. Willingness is emotional and behavioral: how you react to drawdowns, how you sleep, whether you start revenge-trading your own long-term plan. Ability is structural: whether a drawdown would actually change your life.

Those two can point in opposite directions. A young professional with stable income might have high ability-time and cash flow can absorb volatility-but low willingness because a 30% drawdown feels like a personal failure. A retiree with a paid-off house might feel emotionally unbothered by swings (high willingness), yet have low ability because a deep drawdown early in retirement can permanently impair spending. Confusing willingness and ability produces portfolios that look bold on paper and brittle in reality.

Constraints also hide inside the word "need." Need is a sharp knife. If you *need* a certain return to meet a goal, you are tempted to treat that need as a guarantee you can demand from the market. Markets do not negotiate with your needs. If your plan requires 12% annual returns with low drawdowns, the issue isn't whether you're optimistic; it's whether you're pricing reality correctly. A need-based return target can be useful as a diagnostic, but it becomes dangerous when it turns into entitlement.

Here's an uncomfortable but clarifying exercise: write down the maximum loss you could tolerate in dollars, not percentages, over the next year without changing your lifestyle. Many people discover that their "aggressive" self-image collapses when translated into grocery money. A 25% drawdown on a $1,000,000 portfolio is $250,000. That number is either survivable or it isn't. Percentages are how investors talk. Dollars are how investors behave.

The same translation matters for time. Ask: if your portfolio dropped 30% this year, could you still stick to your plan for the next three years without needing the money? If the honest answer is no, then 30% isn't "volatility"—it's a threat to liquidity. And if it's a threat to liquidity, it will eventually become a threat to decision quality, because it will corner you into choices you didn't want to make.

Taxes are another constraint that loves to pretend it's just a detail. Taxes are not just a line item; they are a friction that changes the true cost of turning over a strategy. A short-term trading gain and a long-term capital gain can have very different after-tax outcomes, and the difference is not trivia when compounded over years. A strategy with a modest pre-tax edge can be neutralized by high turnover in a taxable account. Meanwhile, a "boring" strategy that harvests long-term gains can win by doing less.

Leverage deserves a special mention because it is the constraint-killer. Leverage makes everything look better in backtests and feel worse in real life. It compresses time: what would have been a tolerable drawdown becomes a margin call; what would have been a temporary paper loss becomes a forced liquidation. Leverage doesn't just increase risk; it changes the *shape* of your risk by introducing cliffs. If your constraints include "I can't blow up," leverage should be treated like a controlled substance, not a productivity hack.

Even without explicit leverage, many investors have hidden leverage through concentrated positions. Concentration can be rational-some fortunes are built that way-but it is never "free conviction." It is a decision to trade diversification for variance. That trade might be appropriate if you have deep knowledge, long horizon, and the ability to withstand ugly drawdowns without needing to sell. But if concentration is really just boredom with diversification, it's not a strategy; it's a craving.

All of these constraints point to a practical conclusion: you need a personal definition of "winning" that isn't just "make money." Winning might mean "fund college in 2036." It might mean "never risk selling my house." It might mean "grow my capital without needing to check quotes every hour." It might mean "take enough risk to outrun inflation without gambling with essentials." Until winning is defined, decision quality is impossible to judge, because you're grading without a rubric.

That rubric is what an Investment Policy Statement (IPS) is meant to be. The name sounds institutional, like something written by a committee in a mahogany room, but the purpose is refreshingly human: it's a pre-commitment device. It's you, on a calm day, writing instructions for yourself on a stormy day. The IPS doesn't need to be long. It needs to be unambiguous in the places where ambiguity will hurt you most: objectives, constraints, and the rules that keep you from improvising with your future.

A good IPS is not an attempt to predict the market. It's an attempt to prevent *you* from becoming the market's easiest counterparty. In practice, that means codifying things like: your target asset allocation, your rebalancing rules, what kinds of investments are off-limits, how much liquidity you keep outside the portfolio, and the circumstances under which you would change the plan. The most valuable sentence in an IPS is often the one that says, plainly, "I will not change my allocation based on recent performance."

That sentence sounds obvious right up until it becomes hard. Performance-chasing is not stupidity; it's a natural response to social pain. When your neighbor's "simple" portfolio is ripping and yours is lagging, your brain interprets underperformance as status loss. It starts inventing narratives about why your plan is outdated. It starts looking for permission to abandon your discipline. Rebalancing—selling what's gone up and buying what's gone down—then feels like an act of self-harm. The IPS exists to make the disciplined action feel normal by making it pre-decided.

Constraints also protect you from another classic self-trap: mixing time horizons in the same bucket of money. When short-term money and long-term money share an account, the short-term needs tend to win every argument. A market drawdown suddenly feels like a threat to everything at once, even if only a portion of the portfolio was truly time-sensitive. Segregating goals—mentally and sometimes literally via separate accounts—can improve decision quality because it reduces the chance that a near-term need hijacks a long-term plan.

There's a psychological trick here that is worth using: treat different goals as different clients. Your "retirement client" might be able to tolerate volatility. Your "down payment client" cannot. If you'd never put a retiree into a portfolio designed for a house purchase in two years, don't do it to yourself just because it's the same name on the account statement. The market doesn't care which dollar is which, but your constraints do.

One of the most common constraint failures is the "all-weather" fantasy: a portfolio that will deliver strong returns, low volatility, high liquidity, and perfect timing flexibility all at once. That portfolio does not exist. Every portfolio is a set of trade-offs, and trade-offs are not flaws; they're the price of coherence. The real failure is not that your portfolio has weaknesses. The real failure is not knowing what they are until the market finds them for you.

When constraints are explicit, you can start making bets that are not merely intelligent, but *compatible*. Compatibility is underrated. A portfolio that makes sense with your income stability, your spending needs, your temperament, and your time horizon will outperform a theoretically superior portfolio that you can't stick with. The most elegant strategy in the world is worthless if you abandon it at the exact moment it's designed to work.

This compatibility lens also changes how you interpret "risk." Risk is not simply volatility. Volatility is movement. Risk is the chance that movement forces you into a bad decision—selling low, levering up to catch up, or abandoning a well-founded plan because the discomfort became unbearable. If you define risk as "the probability I become my own worst enemy," constraints stop feeling like handcuffs and start feeling like guardrails.

The investor's version of adulthood is accepting that guardrails are not optional. They are the thing that makes higher speeds possible. A plan with clear goals and constraints doesn't limit your opportunity; it limits your self-delusion. It forces you to answer questions that the market will otherwise answer for you at the worst possible time: How much downside can you truly live with? How soon might you need the money? What would make you sell? What would make you double down? What do you do when you feel stupid?

Those are not theoretical questions. They are daily questions, disguised as "market commentary."

When you anchor decisions to goals and constraints, the quality of your choices becomes easier to judge—and, more importantly, easier to repeat. You stop trying to win every day, which is a child's game. You start trying to build a portfolio that can survive your real life, which is the only game that matters.

1.3 Market Mechanics: How Money Actually Grows and Shrinks

A portfolio doesn't grow the way most people talk about growth. It doesn't add. It *multiplies*. That single fact explains a surprising amount of investor heartbreak.

Adding is intuitive: you make 10%, you make another 10%, you've made "20%." Multiplying is what actually happens: you make 10%, then you make 10% on a bigger base. That sounds like a free upgrade—until you meet its evil twin. A loss doesn't subtract from the same base you started with. It shrinks the base you're forced to multiply from next.

The market's math is not trying to insult you, but it will. It will happily turn a string of "pretty good" returns into disappointing long-run growth if the path is jagged. It will also turn a small, boring advantage—like lower fees—into a quietly enormous outcome over time. Money is a drama queen in headlines and a monk in arithmetic: it changes its entire destiny based on tiny daily habits.

Start with the most unpleasant piece of truth in personal finance: losses are heavier than gains. Not emotionally—mathematically. A 50% loss requires a 100% gain just to get back to even. That's not a motivational poster; it's the structure of multiplication. Lose half your capital and you are now trying to double what remains. The market may eventually give you that doubling, but you've forced yourself into a harder game.

This asymmetry is why drawdown management is not "conservative." It's just respectful of multiplication. Investors who treat risk as a nuisance—something to tolerate in exchange for returns—are often misreading the trade. The trade is closer to: you accept volatility, but you pay special attention to large losses because large losses permanently lower the platform from which compounding works.

People often respond to that with the obvious question: *Then why take risk at all?* Because inflation exists, because cash has its own form of slow bleeding, and because long horizons usually require participation in risky assets. The goal isn't to avoid volatility like it's a contaminant. The goal is to avoid volatility-driven *ruin*, and to avoid pretending that "average return" tells the whole story.

Average return is a slippery phrase because it sounds like one thing, when it's actually at least two. The *arithmetic average* is what you get when you add returns and divide by the number of periods. The *geometric average* is the growth rate that would produce the same ending value through compounding. In the real world—where returns compound—the geometric average is closer to what you live with.

This is where volatility shows up with a knife hidden behind its back. Two portfolios can have the same arithmetic average return and wildly different long-run outcomes. A smooth path and a jagged path are not interchangeable. The smooth path lets compounding do its job. The jagged path introduces what's often called *volatility drag*: the idea that variability in returns lowers compounded growth, even if the simple average looks fine.

You can see it with a toy example that feels almost too simple, which is exactly why it's powerful. Imagine a portfolio that goes up 50% one period and down 50% the next. The arithmetic average return is zero: (+50% + -50%)/2 = 0%. Sounds like you broke even. You didn't. Start with $100, go to $150, then lose 50% and you're at $75. The geometric reality is a 25% loss. The average return told you a bedtime story. Compounding told you the truth.

That truth becomes more important the more you trade, the more you lever, and the more you use instruments with built-in path dependence. High volatility isn't just "bumpier." It can actually change the destination even when you end at the same place. That sounds mystical until you remember: multiplication cares about sequence. The order of returns matters because the base keeps changing.

This is one reason "just look at the long-term chart" can be good advice and terrible advice. It's good advice when it discourages panic-selling based on short-term noise. It's terrible advice when it encourages people to ignore the path they would actually have to survive. A long-term chart is a postcard. Your real experience is the weather.

Now add friction. Friction is everything that quietly takes a bite out of returns: fund expense ratios, management fees, bid-ask spreads, commissions (when they exist), market impact for larger orders, taxes, margin interest, platform fees, and the many ways "free" trading gets monetized. Friction matters because it is relentless. Volatility comes in waves. Fees arrive like gravity: every day, forever.

The most dangerous thing about friction is that it looks small in isolation. An expense ratio difference of 0.50% versus 0.05% sounds like a rounding error. Over a single year, it nearly is. Over a decade or two, it can be the difference between "retirement works" and "retirement is tight." Fees don't just reduce returns; they reduce the capital that can compound in the future. They are a tax on the future that you agree to in the present, often without feeling it.

There's also a subtle psychological effect: friction changes what "break-even" means. A high-turnover strategy needs a higher edge to overcome costs. That pushes people toward either

- more leverage and more activity to "make it worth it,"
- performance-chasing because the strategy feels like it's always working hard but not getting anywhere.

When you account for friction honestly, many "exciting" ideas downgrade themselves to hobby status.

This is why low-cost investing has such stubborn appeal: it's one of the few advantages you can lock in without needing to be smarter than the market. You don't have to predict anything to pay less. You just have to decide not to overpay. The market can take your forecasts and laugh at them. It can't argue with math that you refuse to donate.

Transaction costs are the friction you notice. Expense ratios are the friction you forget. Taxes are the friction you argue with. All three share a feature: they are certain, while your returns are not. That certainty should change how you think about them. If your expected edge is modest and your costs are non-trivial, you are running uphill on purpose.

Then there's the most misunderstood form of friction: *compounding itself* when applied to leveraged products. Leverage feels like a simple multiplier: *"If the index goes up 1%, I go up 2%."* Over a single day, many leveraged ETFs aim to do something close to that. Over longer periods, the story becomes messy, because the leverage typically *resets daily*. Daily reset means your multi-day return depends on the path, not just the start and end points.

This isn't a conspiracy; it's mechanics. If a product targets 2× the daily return of an index, it must rebalance its exposure as the index moves. In volatile, back-and-forth markets, that rebalancing can turn into a "buy high, sell low" treadmill. Over time, the product can underperform what a naive investor expects from "2x," even if the underlying index is flat or even modestly up.

A quick mental picture helps. Suppose an index drops 10% one day and rises 11.11% the next day. The index is back to even: $100 to $90 to $100. A 2× daily product drops about 20% (to $80) and then rises about 22.22% (to about $97.78). Same start and end for the index, but the leveraged product ends lower. The difference isn't "tracking error" in the casual sense. It's the compounding of leveraged daily returns on a changing base.

The practical lesson is not "never use leveraged ETFs." The lesson is: don't confuse an instrument's *stated objective* with your *intended holding period*. When the objective is daily, but you hold for weeks, you are no longer making the bet you think you're making. You're making a bet on the path. Path bets can be valid. They should be intentional.

That intentionality is where decision quality returns to the conversation, because the math is only half the story. The other half is sizing and survival. A volatile strategy with high expected returns can still be a bad idea if it has a meaningful probability of blowing you up before the long-run edge shows up. This is where growth-optimal thinking-often associated with the Kelly criterion-gets both celebrated and abused.

The Kelly intuition is simple and seductive: if you have an edge and you're making repeated bets, there is a bet size that maximizes long-run compounded growth. It's the strategy that, in theory, turns advantage into the fastest exponential curve. The seduction is that it sounds like a universal instruction: "maximize geometric growth." The reality is messier, because real investors are not frictionless log-utility robots with perfect knowledge of the odds.

In the real world, you don't know your edge precisely. You estimate it, often with overconfidence. Your return distribution is not stable.

Your correlations change under stress. Your liquidity can vanish. Your personal constraints can change. In that environment, the "optimal" bet size is an illusion with sharp edges. Overbetting doesn't just reduce growth; it increases the chance of catastrophic drawdowns that you cannot psychologically or financially recover from.

So the grown-up version of growth-optimal thinking is not "bet Kelly." It's: understand that compounding punishes over-sizing, and then size in a way that respects estimation error. Many professionals effectively "fractional Kelly" without using the name: they scale down because they know the numbers are not as certain as the formula pretends. The point isn't to worship a formula. The point is to internalize the shape of the risk: small edges do not justify large bets when the left tail is fat and your inputs are shaky.

Volatility drag, friction, and leverage mechanics all point to the same theme: the market rewards not just intelligence, but *robustness*. Robustness is the ability to keep functioning when your assumptions are slightly wrong, when the path is uglier than you imagined, when costs are higher than you modeled, when you're tired and tempted and your brain wants to simplify the world into "I need to make it back."

One of the most common investor mistakes is treating the portfolio like a collection of independent ideas. "This is my growth sleeve." "This is my dividend sleeve." "This is my trading account." In reality, everything is connected through your balance sheet and your nervous system. A drawdown in one bucket can force liquidation in another. A margin call doesn't care that the trade was "separate." Your constraints don't either. The mechanics of money are not modular.

That's why the humble question "What is my actual compounded return after all costs?" is so powerful. It is an antidote to self-deception. It forces you to confront whether your activity is producing net progress or just producing motion. Motion feels like work. Compounding is the result. Investors routinely confuse the two.

There's also a subtle but important distinction between *risk you choose* and *risk you inherit*. You choose risk when you buy equities because you believe in long-run growth. You inherit risk when you pay high fees, trade too often, ignore taxes, use instruments with hidden path dependence, or size a position as if your estimate of edge is gospel. The chosen risks can be part of a coherent plan. The inherited risks are often just negligence.

None of this requires advanced mathematics. It requires respect for the mathematics you already live in. Compounding is not a feature you optionally turn on; it is the operating system. Volatility is not just a feeling; it is a variable that changes long-run growth. Friction is not "small"; it is a permanent claim on your future. And leverage is not a multiplier; it is a reshaping of your distribution, usually in the direction of fragility.

If you want a practical north star, it's this: focus less on finding the trade that makes a great story, and more on building a return path you can actually survive. The market doesn't grade you on your thesis statements. It grades you on what your capital becomes after the math has had its say.

And the math always has its say.

Chapter 2

The Math of Beliefs: Probabilities and Payoffs

Uncertainty isn't an enemy to be eliminated; it's a resource to be managed. While amateurs look for guarantees and absolute certainties, professionals hunt for mispriced odds. This chapter hands you the mathematical flashlight needed to navigate the dark room of market randomness, helping you distinguish between a bad decision and a merely bad outcome by focusing on process over results.

2.1 Thinking in Probabilities: Becoming a Better Forecaster

A dangerous sentence in markets isn't "I don't know." It's "I'm sure." Certainty feels like competence. It has a clean, satisfying click to it, like the lid of a well-made watch case. But markets don't reward the *feeling* of being right. They reward being right *often enough, early enough,* and with a position sized to survive the times you aren't. The most reliable upgrade you can make to your decision-making isn't a new indicator or a better news feed. It's swapping binary declarations—*buy or don't buy, risk-on or risk-off, recession or no recession*—for probabilities that admit what reality looks like: fuzzy edges, partial information, and outcomes that can be "wrong" without being stupid.

Probabilities sound like math class, but they're closer to good conversation. When someone says, "I'm 70% sure," they aren't being evasive; they're being specific about uncertainty. They're acknowledging two truths at once: *I have reasons*, and *those reasons are not the world*. That second clause is where most accounts go to die. People don't blow up because they lack intelligence. They blow up because their confidence runs hotter than their evidence, and because the market is an unforgiving auditor of overconfidence.

Binary thinking sneaks in because it's efficient. The brain wants to conserve energy, and nothing is cheaper than "yes/no." It's why headlines work. It's why pundits talk in absolutes. It's why your own mind can look at three bullish datapoints and declare a new era. A probability is cognitively more expensive: it forces you to ask, "How often does this sort of story end the way I'm imagining?" and then, worse, it forces you to put a number on it. The number is a commitment. A number can be graded. A number can embarrass you in your own journal six weeks later.

That grading is the point. In markets, you don't get to be judged by the elegance of your explanation; you get judged by whether your bets cash. But there's a deeper discipline than P&L: learning to judge your *beliefs* the way a casino judges games. Casinos don't need to predict which exact roulette spin lands on black. They need the odds to be right. They need the probabilities to be calibrated so that, over time, the edge shows up. Investors and traders often try to act like prophets (calling the spin) when they should be acting like bookmakers (setting the line).

Calibration is the unglamorous superpower behind that mindset. A forecaster is calibrated when "70%" means the event happens about seven times out of ten over many similar calls. It's not about being confident; it's about being *appropriately* confident. If your "90% sure" predictions come true only 60% of the time, your model of the world is lying to you, and the lie is wearing a tuxedo.

A useful way to feel calibration in your bones is to imagine two friends. One friend speaks in timid probabilities: "I'm 55% sure," "maybe 60%." The other speaks like a hammer: "Guaranteed," "no doubt," "lock." The timid friend may sound less impressive at dinner, but in the long run they are often closer to the truth, because they're leaving room for what they don't know. The hammer friend may be right spectacularly on some calls, and that occasional spectacle is exactly what makes them dangerous. It trains everyone around them to confuse *style* with *skill*.

Markets turbocharge that confusion. A confident story that happens to work once can buy a lifetime of credibility. A cautious, probabilistic call that's correct can be dismissed as "lucky" or "too hedged." This creates a perverse incentive: make bold claims, attract attention, and let the audience forget the misses. If you want to become a better forecaster, you need a private standard of excellence that doesn't depend on applause. That standard is simple: *when I say 70%, do I mean 70%?*

The first step is learning to stop speaking in disguised certainties. Many sentences that sound probabilistic are just certainty wearing a wig. "I think it's likely" can mean 55% or 95% depending on the mood. "There's a good chance" often means "I want to buy, but I'd like plausible deniability." Even "high conviction" is usually an emotion, not a number. If you can't attach a probability, you don't have a forecast; you have a vibe.

Attaching a number feels artificial at first, because it exposes the machinery of your thinking. You'll notice, quickly, that you're guessing. That's fine. The goal isn't to summon perfect probabilities from the void; it's to make your uncertainty explicit so it can be improved. Think of it like weighing yourself when you start training. The first measurement isn't there to shame you. It's there so that change is measurable.

A practical way to begin is to force your brain into ranges. Instead of "rates will be cut," try: "By the end of the next two meetings, I think a cut is 25%." Instead of "this earnings report will be good," try: "I think the stock closes up at least 3% tomorrow with 40% probability." Notice what happens immediately: you have to define the event precisely. You have to define the time window. You have to define what "good" means in dollars, not adjectives. Probabilities don't just make you humbler; they make you clearer.

Clarity also makes you confront base rates—the outside view, the statistician's antidote to "this time is different." A base rate is simply how often something happens in a relevant reference class. How often do companies that miss revenue by 5% guide up? How often do high-multiple software stocks recover within a month after a 20% gap down on earnings? How often does an initial breakout above a prior high fail within two weeks in this volatility regime? These questions are not poetic. They are profitable.

The inside view is the story you're tempted to tell: this company is special, this management team is different, this product cycle is unique, this macro setup is unprecedented. Sometimes that story is true. Often it's just your brain trying to protect its favorite narrative from the brutal boredom of statistics. The base rate doesn't prevent insight; it prevents delusion. It forces you to anchor your forecast in reality before you let yourself decorate it with cleverness.

Here's the uncomfortable part: base rates are frequently lower than your intuition. Human beings are built to detect patterns, and we often mistake pattern detection for inevitability. You see a chart that looks like a coiled spring, and you *feel* the breakout. You hear a charismatic CEO describe a new market, and you *feel* the S-curve. You watch inflation print cool for two months, and you *feel* the pivot. The base rate asks, coldly: "In similar setups, how often does that feeling get paid?" Feelings are data. They are just not decisive data.

A base rate also does something psychologically clever: it gives you a default position. Without it, you are forced to invent a probability from scratch, and your invented number will be dragged around by whatever you saw most recently. With it, you start with a grounded prior belief and then adjust as evidence arrives. That is what good forecasters do in every field that actually grades them. Weather forecasters don't wake up and declare "90% chance of rain" because the sky looks moody. They begin with models and historical patterns, then incorporate new information. Investing is messier than meteorology, but the habit is transferable: start with the outside view, then earn your deviations.

Earning your deviations is where many investors subtly cheat. They don't consciously lie; they just grant themselves too much permission to say "but this case is special." Special cases exist. But you should treat "special" like a scarce resource. If every trade is a special case, you don't have a process; you have a collection of excuses. A good discipline is to demand a written reason for every base-rate override, and to demand that the reason be *testable*. "This management team is visionary" is not testable. "This company has net revenue retention above 130% for eight consecutive quarters, and the peer group with this metric has historically re-rated upward after profitability inflects" is at least something you can later check for truth.

Notice what's happening: you're turning a narrative into a claim. Claims can be wrong in a useful way. When you track them, you learn. When you don't, you just recycle them with the confidence dial turned up.

Tracking is the hidden engine of better forecasting. Most investors remember their wins like trophies and their losses like administrative errors. Memory is a biased accountant. The cure is a forecast journal that records probabilities and definitions before the outcome is known. It doesn't need to be elaborate. It needs to be honest. Date it. Specify the event. Assign a probability. Record the reasons in a few bullet points. Then, later, record what happened.

This is where many people quit, because they confuse journaling with self-punishment. It doesn't have to be. Think of it as running a personal research lab. You're not tracking to prove you're smart; you're tracking to locate which parts of your thinking are miscalibrated. Maybe your 80% calls are fine, but your 60% calls are actually coin flips. Maybe you're excellent on single-stock earnings and terrible on macro. Maybe your conviction is reliable only when it's grounded in valuation and not when it's grounded in narrative momentum. Without data on yourself, you're just guessing about your own strengths.

Once you have enough resolved forecasts, you can score them. One common scoring method is the Brier score, which punishes overconfidence brutally. It's simple: take your predicted probability, compare it to what happened (1 for yes, 0 for no), square the difference, and average over time. A 90% forecast that's wrong is not "almost right." It's a large error. This is exactly the emotional lesson most investors need. Overconfidence isn't a personality quirk; it's a measurable leak.

Calibration also changes the way you handle "near misses." Suppose you forecast a 70% chance that a stock is up on earnings, and it's down 1%. The impulse is to say, "I was wrong." The more accurate statement is: "A 30% outcome occurred." That subtle shift matters. It prevents you from thrashing your beliefs after normal variance. It also prevents you from getting cocky after a win that was always supposed to happen only 55% of the time. Over a long enough series, you learn to feel the difference between bad luck and bad forecasting. That is emotional maturity disguised as mathematics.

There's a second upgrade that comes with probabilities: you begin to see that a forecast is not a single number. It's a distribution—a spread of plausible outcomes with different weights. Binary thinking makes you focus on the center of that distribution, the "most likely" story. Probability thinking forces you to look at the tails: the outcomes that are less likely but far more consequential. In markets, tails are where careers are made and accounts are destroyed. You don't need to predict tails perfectly. You need to respect them enough to price them into your decisions.

Respecting tails often starts with a question that sounds almost childish: "What else could happen?" Most people ask it as an afterthought, like tapping the brakes at the bottom of a hill. A good forecaster asks it early, and then assigns probabilities to the alternatives. Not as a theatrical exercise in humility, but as a real attempt to map the space of outcomes. If you're bullish a semiconductor stock, the alternatives are not just "up less than I think." The alternatives might include: a competitor announces an architectural leap, an export restriction changes the TAM, a key customer in-sources, a cyclical downturn hits orders faster than the market expects. Each one may be low probability. Together, they form a meaningful chunk of reality.

Base rates help here too. If you know that a certain industry has a long history of sudden pricing wars, you don't get to treat pricing as a footnote. If you know that leverage turns small misses into equity disasters, you don't get to ignore balance sheets because the story is exciting. Base rates are not only about whether an event happens; they're about what kinds of events are *common* in the ecosystem you're playing in.

An underappreciated base rate in investing is that most explanations are post-hoc. After a big move, commentators will confidently explain why it happened, even when the move was driven by positioning, liquidity, or simple surprise. The base rate of "clean causality" is low. If you train yourself to assign probabilities, you'll start to notice how many market narratives are just certainty theater. That awareness makes you harder to manipulate and less likely to chase.

Another underappreciated base rate is about you. How often are *you* right when you feel a certain way? Do you tend to buy breakouts that fail? Do you tend to fade momentum too early because you hate paying up? Do you tend to get bearish after a drawdown because the pain hijacks your time horizon? These are not moral failings. They are predictable patterns in human behavior. But if you don't measure them, you'll keep calling them "intuition."

This is where probability thinking becomes a mirror. You'll notice which emotions inflate your confidence. Anger is a classic one: the market "shouldn't" be doing this, therefore it must revert. Another is pride: you found the idea early, therefore it must be good. Another is fear of missing out: everyone is making money except you, therefore the trade is "obvious." Probabilities don't remove these emotions. They force you to translate them into a number, and that translation often reveals how thin the evidence really is.

To make the translation easier, it helps to build a personal probability vocabulary. "Coin flip" should mean something like 50–55%. "Slight edge" might mean 55–60%. "Meaningful edge" might mean 60–70%. "Strong" might mean 70–80%. Reserve 90% for rare situations where the structure of the world is on your side: arbitrage-like constraints, contractual cash flows with limited uncertainty, or very short-horizon events where information is unusually complete. If you find yourself living at 85–95% all the time, you're not confident; you're uncalibrated.

A good test for whether you deserve a high probability is whether you can comfortably defend the opposite side. If you can't articulate a plausible bear case, your probability is contaminated by identity. You're not forecasting; you're campaigning. Markets are not a debate club where the best speaker wins. They're a weighing machine for incentives, constraints, and surprise. Surprise is what hurts, and surprise lives in the arguments you refused to take seriously.

Base rates also discipline your time horizon, which is where many forecasts cheat. A bullish long-term thesis can be true while the stock gets crushed for a year. A bearish macro view can be true while markets melt up for six months. If you don't specify *when*, your forecast becomes unfalsifiable. Unfalsifiable forecasts feel safe. They are intellectual junk food. A probability attached to a date forces you to stop hiding behind vague timelines.

When you get serious about dates, something else becomes obvious: your forecasts have different half-lives. Some beliefs decay quickly. A chart-based setup might be obsolete in three days. A fundamental thesis might take three quarters to play out. A macro regime call might require you to watch monthly or quarterly indicators. Without stating the half-life, you'll either cling to stale information or abandon good ideas prematurely. Probability thinking encourages you to attach not only a number but also an *expiration date* to your conviction.

Consider how this changes the way you consume news. Most people treat new information as a vote: bullish or bearish. A better approach is to treat it as evidence that should adjust your probability up or down by a specific amount. If you expected a company to guide conservatively and they guide aggressively, does your probability of "stock up 10% in three months" move from 55% to 65%? Or from 70% to 72%? The size of the update is the heart of good forecasting. Small updates are often more realistic than dramatic flips, because most single datapoints are noisy. Dramatic flips feel decisive, but they are usually a form of emotional relief.

You can also use base rates to avoid being seduced by vivid anecdotes. A charismatic founder might convince you their product is revolutionary. Maybe it is. But the base rate of "revolutionary product → great investment at today's price" is not 100%. Sometimes the product wins and the stock still disappoints because the valuation already assumed perfection. Sometimes execution fails. Sometimes competition copies. Sometimes regulation intervenes. The anecdote is the spark; the base rate is the firebreak.

One of the most liberating realizations for an investor is that a forecast can be good even when it loses money, and it can be bad even when it makes money. If you buy a stock because you're "sure" it will go up tomorrow, and it goes up, your process may still be terrible. If you buy with a well-calibrated 55% edge and it goes down, your process may still be excellent. Probabilistic thinking is what lets you separate decision quality from outcome quality. Without it, every loss feels like stupidity and every win feels like genius. That emotional roller coaster is not just exhausting; it trains you into worse habits.

When you begin to think in probabilities, you also start to notice that many of your "bets" aren't isolated. They share hidden drivers. You might think you have six different ideas, but they're all quietly long liquidity, long growth, or long risk appetite. Your forecasts will be correlated. That means a string of losses isn't necessarily proof that your edge disappeared; it might simply be the world delivering one bad regime for your cluster of beliefs. Probability thinking doesn't solve correlation, but it makes you humble about it. If six bets depend on the same macro tailwind, your effective probability of "a bad week" is higher than your spreadsheet suggests.

A surprisingly effective calibration exercise is to forecast mundane, non-financial events for a while. Will it rain tomorrow? Will a flight be delayed? Will a sports team win? These events have clear resolution and abundant base rates. The point isn't to become a meteorologist. The point is to train your brain to attach numbers and then receive feedback. Most investors operate without feedback loops that are clean and fast. Training calibration on everyday events builds the muscle without the emotional charge of money. Then you bring the habit back to markets, where your ego will fight it.

Ego is the real opponent. The market doesn't care if you were early. It doesn't care if your thesis was sophisticated. It doesn't care if you "basically" got it right. It only cares what happened and how you were positioned when it happened. Probability thinking is a way to keep your ego out of the driver's seat. When you say "I'm 62%," you are admitting, in public or at least on paper, that you can be wrong. That admission is not weakness. It is the price of admission for learning.

Learning, in turn, requires that you treat mistakes as information. If your 70% calls are hitting 50%, you don't fix it by declaring you'll be more confident next time. You fix it by asking where the error came from. Were you using the wrong reference class for your base rate? Were you overweighting a noisy indicator? Were you ignoring a regime shift? Were you confusing a good business with a good price? Each of those errors has a different remedy. Without probabilities, all errors feel the same: "I was wrong." With probabilities, errors become diagnostic.

Diagnostics are where the fun begins. You can discover, for example, that you're systematically overconfident in fast-moving markets and underconfident in slow ones. Or that you're well-calibrated on direction but terrible on magnitude. Or that your edge exists only when volatility is high because your competitors panic, and disappears when markets are calm because everyone can think clearly. These are not abstract insights. They tell you when to press and when to pass.

Passing is an underrated decision. One of the quiet benefits of probabilistic thinking is that it makes "no bet" feel legitimate. If every opinion must become a trade, you'll trade in the fog. If you allow yourself to say, "I'm 52% bullish and the payoff isn't attractive," you'll avoid a lot of low-quality activity that exists mainly to relieve boredom. Boredom is a costly emotion in markets. It sends people looking for action when what they need is selectivity.

Selectivity doesn't mean paralysis. It means that your confidence has to be earned. A good forecaster develops a sense for when a probability is anchored in evidence versus when it's floating on hope. Evidence tends to be repeatable: data, comparable cases, structural constraints, incentives. Hope tends to be unique: a one-off story, a gut feeling, a narrative that can't be tested. Hope can occasionally make you money. Evidence can make you money *repeatedly*. That repetition is what turns a trader into a professional rather than a gambler with a good year.

Probabilities also change how you talk about your positions. Instead of "I'm long because this is going to rip," you start thinking like: "I'm long because I think there's a 65% chance of a positive surprise relative to expectations, and the market is pricing something closer to 50%." That is a very different sentence. It forces you to define what the market is already assuming, which is where many people forget to look. A stock is not cheap because it's down. It's cheap if the price implies a gloomier probability distribution than reality.

This is why base rates should be paired with expectations. The base rate tells you what typically happens. Expectations tell you what is already in the price. Your edge, when it exists, lives in the gap between the two. Without probabilities, you can't see that gap clearly; you just feel it. Feelings are not nothing, but they are not a strategy.

Over time, the goal is not to become a robot. The goal is to become a disciplined human: someone who can have strong views without needing them to be certain, someone who can be wrong without feeling humiliated, someone who can change their mind without treating it as betrayal. The market will still surprise you. That's not a bug; it's the business. The forecaster's job is to be surprised less often than the price implies, and to size those surprises so that when you're wrong—as you will be—you live to make the next forecast.

The final irony is that probabilistic thinking can make you more decisive, not less. Binary thinkers hesitate because they secretly demand certainty before acting. They wait for the "confirmation" that never arrives until after the move. Probabilistic thinkers act when the odds and payoffs justify action, even while admitting uncertainty. They can say, calmly, "I don't know for sure, but I believe the odds are in my favor." That calm is not bravado. It's calibration.

And calibration, unlike swagger, compounds.

2.2 Expected Value: The Engine of Decision Making

A probability without a payoff is just trivia. You can be exquisitely calibrated about whether a coin lands heads and still be broke, because no one pays you for being right about coins. Investing is different only in the details: the market doesn't pay you for accuracy; it pays you for *edges* applied to *payoffs*. Expected value is the bridge between "I think this will happen" and "I should risk money on it." It turns a forecast into a decision.

Expected value (usually shortened to EV) is the probability-weighted average of outcomes. If an outcome can happen in a few distinct ways, EV is computed by multiplying each payoff by its probability and adding the results. In plain language: imagine every possible future as a line item on an invoice; EV is what you'd expect to pay or receive if you could run that invoice a very large number of times.

The simplicity is deceptive. EV looks like a tidy formula, but it's an attitude toward life: *stop asking whether you'll be right, and start asking what "right" pays and what "wrong" costs.* Most investing mistakes aren't math mistakes. They're category mistakes. People buy trades that feel likely but pay little, and avoid trades that feel uncomfortable even when the payout is enormous. EV forces those two instincts to meet in the same room.

A quick numerical example makes the point sharper than any metaphor. Suppose you have a trade idea with two outcomes over the next week:

- 60% chance you make \$100
- 40% chance you lose \$200

The EV is $0.60 \times 100 + 0.40 \times (-200) = 60 - 80 = -20$. You might win *most* of the time and still have a negative game. Casinos love this structure; they just reverse the signs. Many retail traders unknowingly volunteer to be the casino's customer by focusing on win rate instead of EV.

Now flip it:

- 40% chance you make \$400
- 60% chance you lose \$100

EV is $0.40 \times 400 + 0.60 \times (-100) = 160 - 60 = +100$. You lose *most* of the time and still have a positive game. This is the emotional hurdle: your ego hates a strategy that loses 6 times out of 10, even if your bank account loves it.

EV also reveals why "being right" is a flimsy identity in markets. A trader can be right 70% of the time with tiny wins and occasional disasters and still end up as a cautionary tale. Another can be "wrong" frequently but occasionally catch a trend, a volatility spike, or an asymmetric option payoff and quietly build a fortune. The difference is not virtue. It's distribution.

A distribution is just the shape of your outcomes: how often you make small money, lose small money, and experience rare windfalls or rare blow-ups. Two strategies can share the same average outcome and feel completely different to live through. One might be smooth, dull, and quietly compounding. Another might be long stretches of small losses punctuated by a few spectacular wins. EV is the average; the distribution is the biography.

Right after traders learn EV, many of them make the same mistake: they treat it like a stamp of approval. "Positive EV" becomes a permission slip to take any bet that looks mathematically attractive. That's like saying "this restaurant has good average reviews" and ignoring the fact that half the diners got food poisoning. EV is necessary, but it is not sufficient. Your ability to *survive* the distribution matters. A strategy with a wonderful EV can still be untradeable if it routinely puts you near ruin.

Still, EV is where discipline begins, because it forces you to write down the game you're actually playing. Most investors carry an implicit EV model without realizing it. They'll say, "I like this stock" while secretly assuming, "the upside is 30% and the downside is maybe 10%." They'll say, "this is safe" while secretly assuming, "the downside is 2%." The market's job is to show them that their secret assumptions were fantasies. Writing the assumptions down doesn't guarantee correctness; it guarantees visibility.

There's also a more subtle benefit: EV makes you confront what you're optimizing. Are you optimizing for expected dollars over one trade? Expected dollars over a year? Expected *utility* (how you emotionally experience gains and losses)? Or expected *log-growth* (the compounding rate of your wealth over repeated bets)? These goals can disagree, sometimes violently. A bet that maximizes expected dollars might be so volatile that you can't hold it, or so levered that a bad streak wipes you out before the positive expectation has time to assert itself.

That last point is the reason professional gamblers and some quantitative traders talk about the *Kelly criterion* without necessarily following it to the letter. The Kelly idea, in spirit, says: if you can repeatedly take a bet with a known edge, there is an optimal fraction of your bankroll to risk that maximizes long-run compounded growth (technically, expected log wealth). The Kelly fraction can be uncomfortably large. Full-Kelly sizing often produces drawdowns that are technically "optimal" and psychologically unbearable. The practical takeaway isn't "always bet Kelly." It's that sizing is part of the EV conversation, not an afterthought.

EV also has a cousin in options, where the connection between probability and payoff becomes almost embarrassingly literal. A call option is basically a bet on a range of future prices: if the underlying ends up above the strike, you get paid; if it doesn't, your option can expire worthless. In introductory option pricing, you can model a few possible future prices, assign probabilities, compute the payoff in each scenario, and add them up. That probability-weighted payoff is the option's theoretical value in that toy model. Options make EV visible because the payoff function is shaped like a ramp: limited loss, potentially large gain. That ramp is why options are seductive, and also why they can be strategically powerful. The ramp creates skew.

Skew is the asymmetry of outcomes. Positive skew means you have small, frequent losses and occasional large wins. Negative skew means you have small, frequent wins and occasional catastrophic losses. Many strategies that feel "safe" are secretly negative-skew machines: collecting small premiums until a regime shift arrives like a wrecking ball. Many strategies that feel "painful" are positive-skew machines: taking small paper cuts in exchange for the right to occasionally catch a tsunami.

This is the moment when win rate becomes a dangerous metric. Win rate is emotionally persuasive because it flatters the part of your brain that wants to be right. But win rate is only one coordinate in a larger map. A 90% win-rate strategy can be a negative EV disaster if the 10% losses are huge. A 35% win-rate strategy can be a goldmine if the winners dwarf the losers.

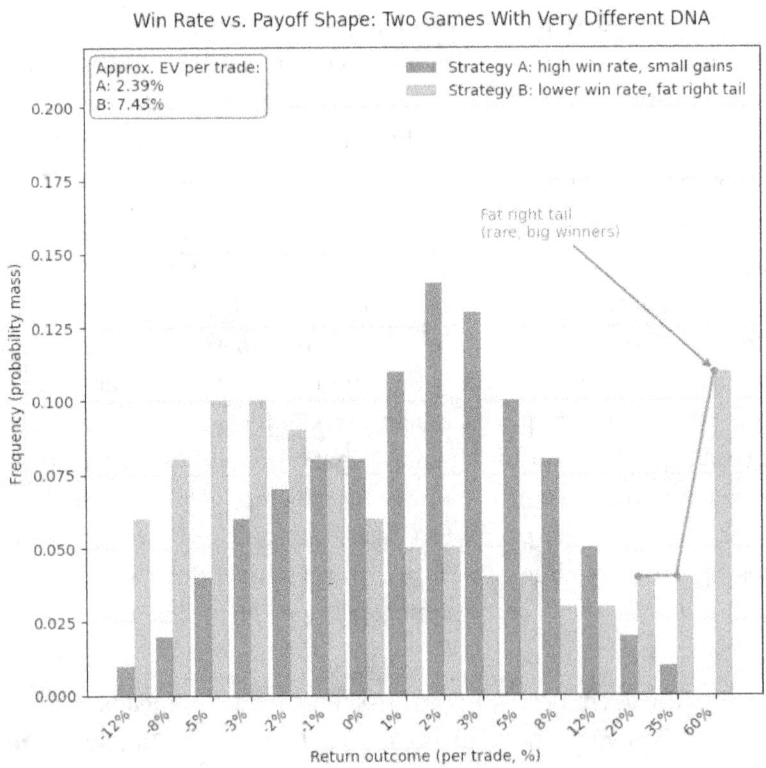

The picture above is a polite way of saying: your brain is going to prefer Strategy A even if Strategy B makes more money. Strategy A showers you with little dopamine pellets. Strategy B makes you feel incompetent for long stretches and then suddenly looks like genius. If you've ever watched traders abandon a good trend-following system after a streak of losses, you've seen this movie. They didn't abandon EV; they abandoned the emotional experience of the distribution.

There's a deeper trap here: people often *mis-estimate* the distribution because they sample the wrong period. In calm markets, the tail looks like a myth. In crisis markets, the tail looks like the only thing that matters. If you build your expectations from a short window, you're not estimating a distribution; you're taking a mood ring reading of the current regime. EV doesn't forgive that. EV calculated on a biased sample is a confident number built on sand.

So how do you actually build an EV estimate in the real world, where outcomes aren't neatly discrete and probabilities aren't printed on the label?

Start with scenarios, not certainties. Scenario thinking is probability thinking with more texture. For a stock, your scenarios might be "beat and raise," "in-line," "miss and guide down," plus a regime scenario like "market risk-off regardless of fundamentals." For each scenario, you estimate a payoff (how the price might respond) and a probability. You can keep the math simple; the discipline is what matters. If you can't describe your scenario set without sounding like you're guessing wildly, that's a signal that you're trading vibes.

Next, make the payoff honest. Payoffs aren't just price moves; they're *position outcomes*, which depend on sizing, leverage, path, and exits. If you say "downside is 10%", but you plan to average down twice and you're using leverage, your downside isn't 10%. It's whatever your behavior makes it. EV is merciless about self-deception. A realistic EV model includes your likely mistakes: slippage, poor fills, late exits, panic sells, revenge trades. If you can't incorporate those, you'll end up with an EV that describes a fictional trader: you, but without your nervous system.

Then apply a discount for uncertainty in your probabilities. This is the investor's version of "measure twice, cut once." If you think your edge is 60/40, but your evidence is thin, treat it like 55/45 until you earn confidence through repetition. Many blow-ups happen not because someone misread the market, but because they overestimated the precision of their own estimates. A probability is not a fact; it's a belief with a number attached. Treating it as a fact is how people justify oversized bets.

This is also where calibration pays rent. If you've trained yourself to assign probabilities that match reality over time, your EV estimates become more useful. Not perfect—markets evolve—but less delusional. Calibration improves the probability side. Experience improves your payoff intuition: how far things can move, how correlations spike, how liquidity disappears, how volatility clusters. EV is where those two kinds of wisdom meet.

A common misunderstanding is that EV tells you what will happen. It doesn't. It tells you which decisions are attractive *over a series*. That "over a series" clause is not legal fine print; it's the entire contract. If you take one positive-EV bet, you can lose. If you take a hundred independent positive-EV bets, you're more likely to win. But independence is rare in markets. Your bets often share a regime driver. That's why EV thinking must be married to risk management: concentration limits, stress tests, and a brutal respect for drawdowns.

Drawdowns are not just financial; they are behavioral. A strategy can have strong EV but lose you money because you can't stick with it long enough for the edge to show up. This is where expected utility sneaks in. A 20% drawdown might be "tolerable" in a spreadsheet and intolerable in a human life. If you optimize for expected dollars and ignore your own psychological constraints, you are optimizing for a trader who does not exist. The better question is: what is the best EV you can *actually* harvest, given that you are a person with fear, pride, rent, and a finite capacity for pain?

This question changes the way you interpret the Kelly intuition. Full-Kelly can be optimal in theory for long-run growth, but it can also be a drawdown factory. Many professionals use "fractional Kelly" sizing (half-Kelly, quarter-Kelly) as a compromise: still responsive to edge, but less likely to produce the kind of volatility that causes self-sabotage. The point is not the exact fraction. The point is acknowledging that sizing is part of decision quality. You can have a great idea and still make it a bad decision by betting too big.

EV also reshapes how you think about "small edges." A tiny EV advantage can be extremely valuable if you can repeat it many times with low risk of ruin—think market making, short-term arbitrage-like strategies, or systematic rebalancing where you're paid for providing liquidity. Conversely, a large apparent EV advantage can be worthless if it requires leverage, illiquidity, or a behavioral tolerance you don't have. The market is full of theoretical edges that nobody can monetize at size. A real edge is one you can *carry*.

There's a beautiful, almost mischievous implication here: sometimes the best EV comes from bets that are *socially uncomfortable*. If everyone is crowding into the same trade, the payoff distribution often becomes negatively skewed: small gains while the crowd keeps buying, then a sharp unwind when the marginal buyer disappears. The trade feels "safe" because everyone agrees. The distribution is quietly sharpening a knife. EV thinking makes you ask, "What happens if the crowd is wrong, and how bad is it?" That question is the beginning of avoiding crowded disasters.

EV thinking also makes you respect boredom. Many of the highest-quality EV opportunities are boring: buying a diversified basket when expected returns are attractive, rebalancing mechanically, collecting a premium for providing liquidity in a disciplined way, holding cash when no good bets exist. Boring doesn't mean easy. It means the story won't entertain you while you wait. If you require entertainment, you'll gravitate toward trades with exciting narratives and ugly distributions.

A final subtlety: EV is not always about maximizing it. Sometimes you knowingly accept a lower EV to reduce tail risk. That's not irrational; it's portfolio thinking. Insurance has negative EV in expectation, and people buy it anyway because it protects against ruin. Hedging can feel like "bleeding premium," right up until the day it saves the business. The mistake is not buying negative EV; the mistake is buying it without understanding that you're buying insurance rather than edge.

When you put all of this together, EV becomes less like a formula and more like an ethical code for decision-making. It demands that you be explicit about probabilities, explicit about payoffs, honest about tails, and realistic about yourself. It also rewards a kind of emotional adulthood: the willingness to endure being "wrong" frequently when the distribution is on your side, and the willingness to reject flattering high win rates when the tail risk is quietly poisonous.

The market will still tempt you with clean narratives and simple metrics. Win rate will still whisper that it means competence. EV answers with a colder, kinder question: *what game am I playing, and does this game pay?*

2.3 Bayesian Updating: How to Learn As Facts Change

Most investors don't fear being wrong. They fear the moment of *realizing* they were wrong, because it forces a choice: update, or defend. Updating costs ego. Defending costs money. The market is remarkably patient with ignorance and brutally impatient with stubbornness, which is why the most expensive phrase in trading isn't "I missed it." It's "I'm sticking with my thesis."

Bayesian updating is the grown-up version of changing your mind. It's not a personality type or a political identity. It's simply a rule for how to revise beliefs when new evidence arrives. The math looks intimidating mainly because it has the nerve to be precise about something people prefer to keep emotional: *how much* should this new fact move me?

At its core, Bayes' rule says:

$$P(H \mid E) \propto P(H)\, P(E \mid H)$$

The symbols are less important than the attitude. H is a hypothesis—your belief about the world ("this company is a winner," "inflation will fall," "volatility is mispriced"). E is evidence—a new piece of information (earnings results, a policy announcement, a competitor's launch, a credit spread move). $P(H)$ is your prior—what you believed before the evidence. $P(E \mid H)$ is the likelihood—how compatible that evidence is with your hypothesis. The output, $P(H \mid E)$, is the posterior—your updated belief.

This is exactly how good analysts already *want* to think. The Bayes framing just stops you from cheating. Without a framework, people update in whichever direction makes them feel smart: they overweight news that agrees with them, underweight news that contradicts them, and then claim "conviction" when what they mean is "comfort." Bayes' rule forces you to look at both sides: how surprising is this evidence if I'm right, and how surprising is it if I'm wrong?

Investing is a live-fire environment for this habit because evidence arrives in the worst possible order. You don't get a clean lab experiment. You get noisy signals, mixed motives, and headlines engineered for attention. You get information that is directionally true and economically irrelevant. You get information that is economically huge but arrives disguised as a footnote. The point of Bayesian thinking isn't to become omniscient. It's to become *harder to whip around* by the wrong kind of newness.

A prior is where most people get defensive, because "prior" sounds like prejudice. In markets, a prior is just a starting point. It is the baseline odds you assign before you see the next datapoint. If you refuse to have priors, you don't become objective. You become erratic. You let the last thing you read set your beliefs.

A strong prior is not a rigid prior. It's a clear one. Strong means *well-grounded*: based on base rates, history, and the structure of incentives. Strong does not mean "high conviction." Sometimes the strongest prior is humble: "I only have a small edge here." A good prior is less like a proclamation and more like a well-anchored tent. You can move it if the wind changes. You just don't let every gust drag you across the field.

Take a concrete example: you're evaluating whether a company's new product will meaningfully expand its market. The inside-view temptation is to be seduced by a demo and a charismatic narrative. The Bayesian approach is to start with a prior based on a reference class: how often do new products in this category become material revenue drivers within two years? How often do they flop? How often do they succeed technically but fail commercially? Your prior might be something like 30% for "material success," 50% for "moderate contribution," 20% for "mostly noise." That doesn't mean you're pessimistic. It means you're starting from a base rate instead of a hype reel.

Then evidence arrives. Suppose early customer reviews are excellent, but adoption is slower than guided. Is that good or bad? The answer depends on how likely you'd see that pattern under each hypothesis. If truly transformative products often have slow enterprise sales cycles early on, then "excellent reviews + slow adoption" might be very consistent with eventual success. If, instead, successful products in this category tend to show rapid uptake because switching costs are low, then slow adoption is a more serious warning. The evidence is the same; the likelihoods differ. This is why two smart investors can disagree without either being irrational. They're using different mental models for how the world produces evidence.

This is also where investors make a costly mistake: they treat evidence as if it is automatically bullish or bearish. Evidence is rarely signed that neatly. A beat on earnings can be bullish if it signals sustainable demand, or bearish if it signals a one-time pull-forward. A margin decline can be bearish if it signals competitive pressure, or bullish if it signals an investment cycle that improves future growth. Bayesian updating is about interpreting evidence *relative to hypotheses*, not relative to your mood.

A practical Bayesian habit is to ask a slightly annoying question every time new information appears: *"Would I be surprised by this if my thesis were true?"* If the answer is "no," the evidence shouldn't move you much. And then the companion question: *"Would I be surprised by this if my thesis were false?"* If the answer is also "no," then the evidence is probably low-quality. It's noise. It's commentary. It's a headline that feels important but doesn't discriminate between competing realities.

The discriminating power of evidence matters more than its emotional impact. Markets are full of high-emotion, low-information events. A CEO interview can be riveting and almost useless. A small change in a credit spread can be boring and extremely informative. Bayesian updating trains you to prize evidence that actually changes odds.

If you want the most intuitive version of Bayes for investing, use the "odds and multipliers" framing. Instead of computing exact posteriors, think like this:

- Start with odds: "I think it's 2-to-1 against."
- Apply an evidence multiplier: "This datapoint makes the bull case about twice as likely."
- Convert back to a probability if you need to.

You don't need to do the conversion perfectly. You need to avoid the common failure mode: treating each new datapoint as permission to leap from 55% to 90%, or from 60% to "I'm out." The market punishes emotional step functions.

There's a spreadsheet-friendly model that makes Bayesian updating feel almost tactile: the Beta-Binomial update, which is just a fancy name for "count wins and losses with a memory." Imagine you're trying to estimate a probability p, like "the probability this setup produces a profitable trade." A Beta distribution can represent your belief about p as if you had seen a certain number of prior wins and losses. These are sometimes called "pseudo-counts." If you set a prior of $\text{Beta}(\alpha, \beta)$, you can interpret $\alpha - 1$ as prior wins and $\beta - 1$ as prior losses. Each new observation updates α or β.

That pseudo-count interpretation is gold for investors because it answers a question people rarely articulate: *how much should this one new trade teach me?* If your prior effectively represents 100 past observations, one new result shouldn't flip your belief. If your prior represents 5 observations, one new result should move you a lot. This is exactly how experience works in real life. Bayes just gives you a dial.

For example, suppose you're testing a short-term mean reversion strategy. You've run it on historical data and you think the true win probability is around 55%, but you know backtests can be optimistic. You encode that skepticism by choosing a prior that isn't too confident, say $\text{Beta}(11, 9)$, which has a mean of $11/(11 + 9) = 55\%$ but only the weight of about 20 pseudo-trades. After 10 live trades, if you go 7 wins and 3 losses, your posterior becomes $\text{Beta}(18, 12)$ with mean $18/30 = 60\%$. Notice what happened: you didn't declare "the strategy is amazing." You nudged your belief upward in a way that respects sample size. If instead you went 3 wins and 7 losses, you'd have $\text{Beta}(14, 16)$ with mean 46.7%. Again: a meaningful update, not an identity crisis.

The deeper lesson is not the Beta distribution. It's that evidence has *weight*, and weight depends on how many times you've seen similar situations resolve. New investors treat each outcome as a verdict on their intelligence. Bayesian thinkers treat each outcome as one more data point in a long trial.

This helps with another common trap: the urge to "wait for confirmation." Confirmation is often just the market moving in your favor, which feels reassuring and teaches you almost nothing about whether your belief was well-founded. The most educational evidence is often uncomfortable: a drawdown that reveals your position sizing is too aggressive, a competitor's unexpected strength, a macro datapoint that contradicts your narrative. A Bayesian mindset doesn't seek discomfort for sport. It simply recognizes that discomfort often carries information.

One of the most dangerous forms of non-Bayesian thinking in markets is anchoring—becoming emotionally attached to your first number. You buy a stock at $100, it drops to $80, and you mentally frame $100 as "fair value" because it's where you first engaged. That anchor has nothing to do with reality. It's just where your ego started counting. Bayesian updating breaks anchors by forcing you to treat price moves as evidence. Sometimes a drop is evidence your thesis is wrong. Sometimes it's evidence the market is offering you a better payoff distribution. The difference is not the drop. The difference is whether the evidence changes the likelihood of your thesis.

A related trap is "thesis creep." You buy because you believe earnings will accelerate. Earnings don't accelerate, so you pivot to "it's a long-term story." That might be true. It also might be you protecting yourself from admitting error by expanding the time horizon. A Bayesian approach makes you define what evidence would falsify your belief and when you expect to see it. Not because you want to be pessimistic, but because you want to avoid turning investing into a religion.

Another subtle enemy is the assumption of independence. Investors are constantly tempted to treat multiple signals as separate confirmations, stacking confidence like poker chips: the chart looks good, the analyst upgrade happened, social media is bullish, a friend likes it, the CEO sounded confident. But many signals are not independent; they are the same information echoing through different channels. If you treat correlated signals as independent, you will update too aggressively and become overconfident.

This error is everywhere in modern markets because information is networked. One bullish narrative spawns ten bullish "facts" that are really just the narrative wearing different outfits. The Bayesian discipline is to ask: *is this new evidence genuinely new, or is it the same evidence being repeated?* If it's repetition, it shouldn't move your odds much.

Independence mistakes show up clearly in macro. Suppose you believe the economy is weakening. You see weaker retail sales, then a dovish comment from a policymaker, then falling long rates. It's tempting to treat these as three independent supports. But the policymaker may be reacting to the same data you are, and rates may be moving because the market expects that reaction. Your three datapoints are partly the same causal chain. Updating three times can turn modest evidence into unjustified certainty.

There's a more practical way to express the same idea: don't count the same thesis twice. If your evidence stack shares a common source, discount it. Social proof is not a likelihood; it's a pressure.

Bayesian updating also pairs nicely with EV thinking, because an updated belief isn't just a new number—it's a new decision. If your probability of a favorable outcome drops from 65% to 55%, that might not mean "sell immediately." It might mean reduce size, tighten risk, or demand a better price. Beliefs and actions are linked by payoffs. A small probability change can justify a large action if your payoff distribution is fragile. A large probability change might justify only a small action if the trade is already sized conservatively with limited downside.

This is the mature investor's edge: not "I always know," but "I always adjust." Markets are rivers of partial information. You're never stepping into the same river twice, and you're never trading with the same knowledge twice. A Bayesian posture keeps you from treating your thesis as a statue. It's a living object, meant to be reshaped.

A particularly powerful habit is to pre-commit to updates. Before an earnings report, write down how you will change your probability estimates under different outcomes. If revenue beats by 5% but margins compress, what happens to your belief? If guidance is strong but the stock sells off, do you treat that as evidence of hidden bad news or as evidence of crowded positioning? Pre-commitment doesn't make you clairvoyant. It makes you harder to hypnotize in the moment.

The market loves to negotiate with your emotions at exactly the wrong time. A pre-committed Bayesian update is like putting your calm self in charge of your panicked self. That isn't a moral victory. It's an operational one.

None of this requires you to compute posteriors with scientific precision. The most important Bayesian skill is directional: updating the right way by the right amount. Overreacting is as costly as underreacting. Underreacting keeps you trapped in bad trades; overreacting keeps you from harvesting good edges because you keep resetting your beliefs to whatever is loudest today.

A final note on ego, because ego is the real math problem. People treat changing their mind as a personal failure, as if their earlier self must be defended in court. That mentality is fatal in markets, because the market is a court with no mercy and no appeals. Bayesian updating offers a different identity: you are not the person who was right yesterday; you are the person who updates fastest without panicking. You are not loyal to a thesis; you are loyal to the process that keeps your beliefs aligned with reality.

That loyalty is rarer than intelligence. And in markets, rare things are often valuable.

2.4 When Everything Moves Together: Regimes and Correlations

Diversification is sold like a seatbelt: clip in, relax, you're safer now. Then the crash happens and you learn an ugly detail about seatbelts: they don't prevent impact; they change what the impact does to you. In a real market storm, the problem isn't that your "risky" assets go down. You signed up for that. The problem is that your *different* risky assets suddenly start behaving like the same asset, as if a hidden magnet snapped on and pulled everything into one trade: *get me out*.

Correlation is the statistic that describes that magnet. It measures how two return series move together: $+1$ means they move in lock-step, 0 means they're unrelated, and -1 means they move opposite. Investors love correlation when it's low because it promises the closest thing finance offers to magic: less volatility without giving up expected return. Investors hate correlation when it spikes because it reveals the punchline: diversification works best when you don't need it.

A correlation is not a causal explanation. It doesn't tell you why two assets move together, only that they do. That sounds like pedantry until you see how people get wrecked. If you mistake correlation for causation, you'll build portfolios that are "diversified" in a story sense (different industries, different tickers, different narratives) but dangerously concentrated in a factor sense (the same sensitivity to growth, liquidity, energy prices, the dollar, or risk appetite). The market doesn't care that you own ten names. It cares what they do when the oxygen thins.

A clean way to think about correlation is that it's a *behavioral* statistic wearing a math costume. In calm markets, investors have the luxury of preferences. They can differentiate between quality and junk, between strong balance sheets and weak ones, between companies with pricing power and companies with hope. In panic markets, preferences collapse into constraints. Margin calls don't ask whether your thesis is elegant. They ask what you can sell. Risk managers don't ask whether the selloff is "overdone." They ask whether you're within limits. When constraints dominate, correlations rise because everyone is reacting to the same forcing function: liquidity.

That's what people mean when they say "everything correlates to one." The "one" isn't a mystical number. It's the shared driver of survival behavior.

The idea of *regimes* is the antidote to naive correlation thinking. A regime is a market environment with a characteristic pattern of volatility, correlation, and dominant narratives. Regimes don't announce themselves with a press release. They creep in, then suddenly feel obvious in hindsight. In one regime, diversification behaves like a gentle shock absorber. In another, diversification behaves like a decorative sticker.

In a fair-weather regime, correlations between many risk assets can be modest and unstable in a good way: your portfolio's rough edges cancel out. In a storm regime, correlations compress upward: your portfolio's rough edges align. This compression is why portfolios that look robust in backtests can become fragile in real crises. Backtests often average across regimes, turning "sometimes safe, sometimes not" into a comforting middle number.

Averages are soothing. Markets are not obligated to be soothing.

The most emotionally confusing part is that correlation can change faster than your beliefs. You can be diligently updating your thesis on each holding, carefully adjusting probabilities as facts arrive, and still lose money because the portfolio-level behavior has shifted. This is the difference between being right about trees and being wrong about the forest. Portfolio outcomes are driven not only by what each asset does, but by how they do it together.

That "togetherness" can be visualized. In calm conditions, if you plot returns of Asset A versus Asset B, you'll often see a loose cloud: sometimes one is up while the other is down, often both are small moves. In crises, the same plot can look like a tight diagonal line: both assets down hard at the same time, as if attached by a rigid rod.

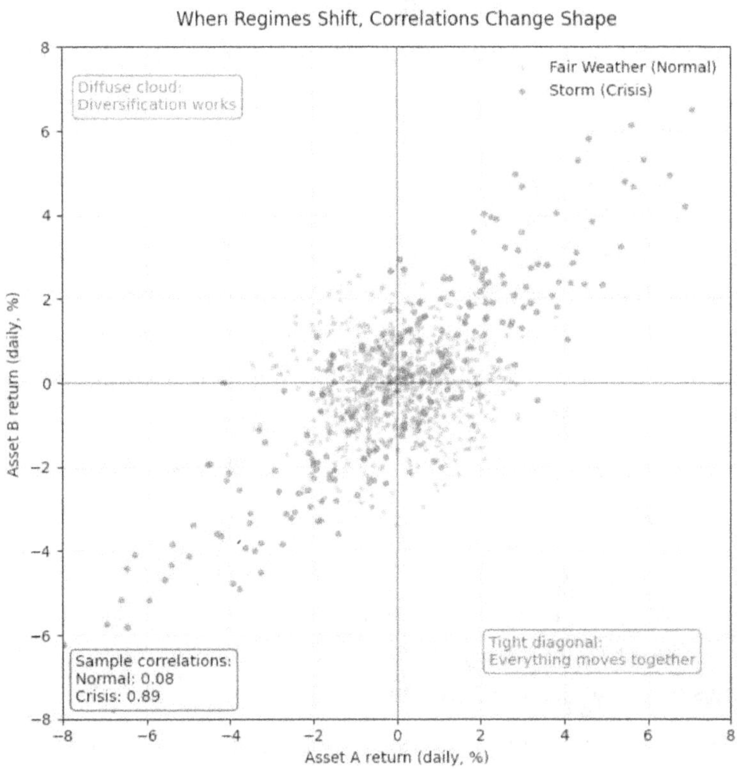

That picture is why "diversify" is good advice and also incomplete advice. The incomplete part is that correlations are not constants. They are *conditional*. They depend on volatility, liquidity, leverage, positioning, and fear. You don't own a fixed correlation matrix; you own a portfolio that gets re-priced by human behavior under stress.

This creates a debate that matters in practice: when we observe correlations rising in crises, is it because linkages between markets truly strengthened (a kind of contagion), or because the measurement gets distorted when volatility explodes? There's a subtle statistical issue here. Correlation is scale-sensitive in weird ways. When volatility rises sharply, simple comparisons of correlations across periods can make co-movement look stronger even if the underlying relationship didn't change as much as it appears. Some researchers have argued that part of "contagion" is really "interdependence plus higher volatility," meaning the world may be less dramatically different than a raw correlation chart implies.

Here's the investing translation: don't treat a crisis correlation spike as a moral revelation about markets ("nothing is diversified!") without asking what's driving the spike. Some of it is mechanical. Some of it is behavioral. Some of it is structural. You don't need to pick one explanation to manage risk well, but you do need to respect that correlation is a noisy witness, especially when the market is screaming.

Even if measurement effects exaggerate the spike, the lived experience remains brutal: diversification benefits tend to shrink during drawdowns, particularly during extreme negative moves. Correlation in the left tail—what happens when things go badly—often matters more than correlation in the middle of the distribution. Two assets can look charmingly diversified most of the year and still fail together when you most need them to offset. Investors don't suffer from average correlation; they suffer from correlation *when they're already losing*.

If you want a more robust intuition than "correlations go to one," think in terms of *shared constraints*. In calm markets, investors can express differentiated views. In stress, everyone faces similar constraints: reduce risk, raise cash, meet margin, comply with VAR limits, satisfy redemption requests. When constraints dominate, assets become connected through the balance sheets of the people who own them. That is not a poetic statement. It is plumbing.

Regimes are how you stop being surprised by the plumbing. A regime perspective asks: what environment are we in, and what environment could arrive quickly? It forces you to treat volatility and correlation as a package deal. Low volatility regimes often coincide with more stable correlations and more reliable diversification. High volatility regimes often coincide with correlation spikes, sudden factor rotations, and a violent repricing of anything that depends on easy financing.

This is where the word *latent* becomes useful. Many regime models treat the market as switching between hidden states—calm and crisis are the cartoon versions—and the state determines the distribution of returns. The state itself isn't directly observed; you infer it from data like volatility, correlations, spreads, and breadth. The important part for a discretionary investor isn't the model. It's the humility: you might think you're in one state while the market is quietly shifting into another.

When people get blindsided, it's rarely because a single stock disappointed. It's because the *state* changed and their whole book was built for the old state.

Correlation is also frequently misunderstood because investors forget it's a *time horizon* concept. Two assets can be uncorrelated intraday and highly correlated over a quarter. They can be positively correlated in bull markets and more correlated (and more negative) in bear markets. They can be diversifying in small moves and not diversifying in large moves. When someone tells you "these are uncorrelated," the correct follow-up is: uncorrelated over what window, in what regime, and for which side of the distribution?

That question is not pedantic; it's the difference between a robust portfolio and a portfolio that looks robust until it doesn't.

Now for the practical part: how do you prepare for regimes and correlations without turning investing into a PhD program?

Start with the uncomfortable audit: list your portfolio exposures in factor language, not ticker language. You don't need fancy tools to begin. Ask simple questions that slice through the stories. How much of your risk depends on falling rates? How much depends on tight credit spreads? How much depends on commodity prices staying calm? How much depends on the dollar weakening? How much depends on "growth gets paid more than value"? If you can't answer, you're not diversified; you're merely distributed.

Then stress-test in narratives you don't like. Investors love to imagine a favorable surprise. They hate imagining a regime shift that makes their favorite assets all move together. But that's exactly the scenario you need to price in. What happens to your "diversified" book if real rates rise quickly? What happens if liquidity vanishes and bid-ask spreads widen? What happens if volatility jumps and your positions become harder to exit at the price you assumed? The point is not to fear every scenario. The point is to notice which scenarios create simultaneous losses across positions.

This is also where the earlier EV mindset matures. Expected value at the position level is not the same as expected value at the portfolio level. A trade can have a good standalone EV and still be a bad addition if it increases your exposure to a crash regime. You don't just want positive EV; you want positive EV *that doesn't all die the same way*.

Rebalancing is the unsexy tool that turns correlation awareness into action. When you rebalance, you restore target weights after markets move. Mechanically, that means selling some of what went up and buying some of what went down. Psychologically, it means doing the opposite of what your nervous system wants. In a fair-weather regime, rebalancing forces discipline and prevents drift into accidental concentration. In a storm regime, rebalancing can be either lifesaving or reckless depending on what's driving the move. If the storm is primarily about temporary dislocation and forced selling, rebalancing can capture a liquidity premium. If the storm is about permanent impairment, rebalancing can be catching a falling knife. Regime thinking helps you distinguish between those two without pretending you can do it perfectly.

Correlation also has a social dimension. In modern markets, crowding can make correlations jump even within a category. If everyone owns the same "quality growth" basket, those names become linked not just by fundamentals but by ownership. When the crowd de-risks, they sell the basket. Correlation rises because portfolios are constructed similarly. This is why diversification based on labels ("tech," "healthcare," "consumer") can fail when the real linkage is "popular with the same institutions."

Another way correlations fool people is through hedges that work in one regime and fail in another. A hedge is not "an asset that goes up when I go down." A hedge is "a position that tends to profit in the specific states of the world that hurt me most." Sometimes bonds hedge equities beautifully. Sometimes inflation or policy shifts make bonds and equities fall together. Sometimes volatility hedges work until the carry cost bleeds you dry and you turn them off right before you need them. A hedge is part statistics, part behavior, part budget. If you can't hold it, it isn't a hedge; it's a story about a hedge.

The right level of sophistication is to avoid brittle certainty. You don't need to forecast the next regime. You need to respect that regimes change and design a portfolio that can tolerate being wrong about the environment. That might mean holding some assets that feel "unnecessary" in good times because they earn their keep in bad times. It might mean limiting leverage even when leverage looks optimal in backtests. It might mean diversifying across strategies (trend, carry, value, quality, volatility) rather than only across tickers. The best diversification is often diversity of *payoff shapes*, not diversity of names.

Correlation is also a reminder that risk is not evenly distributed through time. In many strategies, most of the gains come in calm periods and most of the damage comes in a small number of violent episodes. Those episodes are where correlation and regime awareness matter most. The market can let you feel smart for months and then demand payment in a week. If you've assumed correlations are stable, you'll pay that bill with interest.

A final twist: regimes are not only "calm" and "crisis." There are inflationary regimes, deflationary regimes, tightening regimes, easing regimes, commodity-led regimes, productivity-led regimes. The names matter less than the practical observation: the drivers of returns change, and when drivers change, correlations change. An asset that diversifies you in one macro story may concentrate you in another. Your job is not to find the one true hedge. Your job is to avoid building a portfolio that depends on the world remaining polite.

Polite is not a market condition. It's a phase.

The investor who survives storms isn't the one with the most clever explanations for why correlations "shouldn't" have spiked. It's the one who treated correlation as a living, breathing feature of markets, and treated regimes as a recurring fact of life rather than a once-in-a-century surprise. When everything moves together, it feels unfair. The fairness is irrelevant. The preparation is everything.

Chapter 3

The Human Element: Navigating Biases and Market Ecology

The market isn't a machine; it's a mirror. While we like to imagine charts are driven by cold logic, they are actually powered by millions of humans-and their algorithms-acting on fear, greed, and incomplete information. This chapter peels back the layers of psychology and market structure to reveal who you're really playing against: the crowd, the institutions, and most dangerously, yourself.

3.1 The Mind's Traps: Cognitive Biases and the Narrative Fallacy

Markets don't just test your intelligence. They test your *evaluation system*. The hardest part isn't building a view; it's grading yourself honestly when the tape prints something else. A trader can be thoughtful, risk-aware, and statistically justified—and still lose money today. Another can wing it, oversize it, and win money today. If you wire your brain to learn from *today*, you'll train yourself to do more of whatever happened to work recently, whether it was skill or sunlight.

That's the core psychological trap in trading: the market hands out feedback that looks like coaching but often isn't. In many domains—

surgery, aircraft maintenance, bridge design—bad decisions get punished quickly and consistently. Markets are different. They're a giant stochastic machine: noisy, adaptive, occasionally generous, and more than willing to reward your worst habits for long enough that they become your personality.

Resulting is the name of the first and most expensive mental glitch: judging the quality of a decision by the quality of its outcome. It's not a small bias. It's a complete misgrading of the only thing you actually control.

A simple example: you buy a stock after careful work—clear thesis, identifiable catalyst, valuation margin of safety, and a position size that won't crater you if you're wrong. The next day, the CEO gets hit with an unexpected scandal; the stock gaps down 20%. You did not "make a bad trade." You made a trade with downside you could tolerate, and random life showed up early. If you label that trade "bad," you'll learn the wrong lesson: "Don't do rigorous work," or worse, "Don't take risk at all." The market just trained you to become timid at precisely the moment you should become methodical.

Now flip it: you chase a breakout on pure adrenaline, no model of who's on the other side, no plan for liquidity, and you size it like you're trying to solve your childhood. The stock rips another 8% because a buyback headline hits. You did not "make a great trade." You got paid while doing something fragile. If you label that trade "good," you'll teach yourself that your process is optional, that discipline is for people who don't "have a feel." The market just trained you to repeat the behavior until the one day it doesn't work—and that day tends to arrive in a cluster, not as a polite one-off.

Resulting is seductive because P&L is crisp. It has the satisfying geometry of a scoreboard. But it's also a liar. A single outcome is not evidence of a good or bad decision; it's a sample of one from a distribution you rarely get to observe in full.

Right after a trade closes, your brain wants closure. It wants a verdict. It wants to stamp the decision APPROVED or REJECTED and move on. That impulse is natural and, in markets, dangerous. The more you trade, the more your learning system becomes your edge—or your self-sabotage.

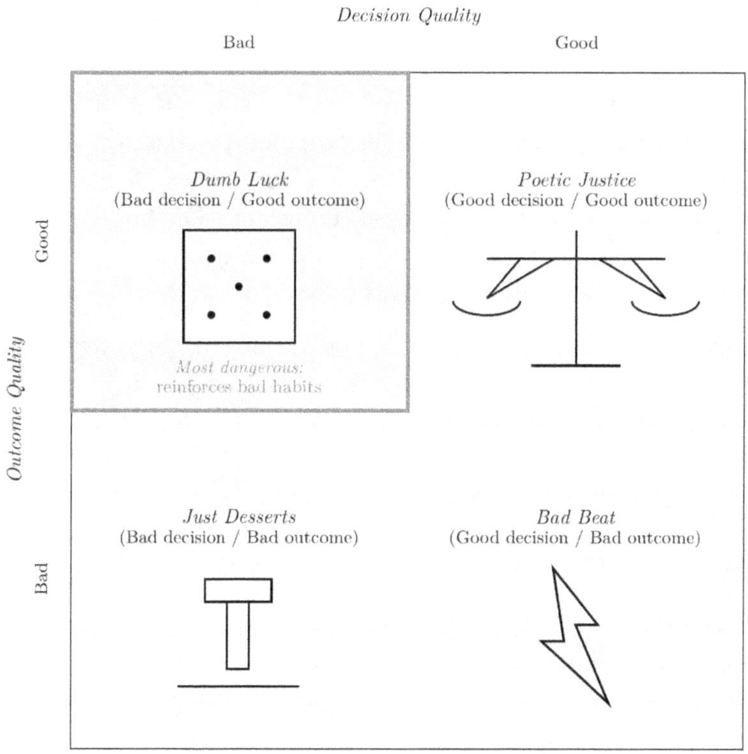

That matrix is more than a cute taxonomy. It's a map of how traders get built—how a process becomes a reflex.

Poetic Justice is the quadrant everyone posts on social media: good work rewarded. *Bad Beat* is where good work gets punished. Those two are psychologically tolerable because they fit a moral story: effort should pay, and sometimes life is unfair. The two that quietly rewire your operating system are the other pair. *Just Desserts* at least gives clean feedback: reckless trade, painful outcome. It's a tuition bill, but the lesson is legible.

Dumb Luck is the slow poison. A bad decision with a good outcome doesn't feel like a warning; it feels like a promotion. It turns randomness into confidence and confidence into leverage. Traders don't blow up because they were wrong once. They blow up because they were wrong repeatedly and got paid anyway—right up until the day the market stopped subsidizing their self-image.

To detox from resulting, the mind needs a different scoring system: one that grades the *inputs*—the information you had, the logic you applied, the risks you accepted—rather than the single realized output. Investors talk about "process" the way people talk about flossing: they know it's good, they don't do it consistently, and they're shocked by the eventual dental bill.

A robust process doesn't mean you never change your mind. It means you know what would make you change your mind *before* you are in the heat of the moment. It means you can say, with a straight face, "This trade lost money and I would take it again," which sounds insane to anyone who confuses decision quality with outcome quality—but is entirely sensible in a probabilistic world.

Here's the unromantic truth: good trades can be losers, and bad trades can be winners. If that sentence makes you uncomfortable, you've located the seam where your psychology will tear under pressure.

The discomfort is not just philosophical; it's biological. Losses don't register as neutral data. They feel like threats. That's not a metaphor. Human beings tend to experience the pain of losses more intensely than the pleasure of equivalent gains, and we tend to judge outcomes relative to a reference point—what we expected, what we "should" have made, where our P&L was an hour ago—rather than in absolute terms. In trading, that reference point can change by the minute, which means your emotional goalposts are constantly moving.

This is why a trader can be up $20,000 on the day, finish up $2,000, and feel like they lost $18,000. The money didn't evaporate; the reference point shifted. Your brain treats "gave back profits" as a loss because it compares the current state to the peak state, not to the starting line. That's the same wiring that makes people feel poor when their neighbor gets rich. In markets, it makes you vulnerable to the two classic sins: taking profits too quickly (to "lock in" the relief of a gain) and holding losers too long (to avoid the crystallization of a loss).

Once that wiring is active, resulting becomes even more toxic. A loss doesn't just hurt; it indicts you. It threatens your competence, your identity, your status at the imaginary tribunal in your head. So you look for exonerating stories. Which brings us to the Narrative Fallacy.

Markets are a factory that produces narratives as a byproduct of price movement. The stories are not merely wrong sometimes; they are structurally incentivized. Financial media must fill airtime. Analysts must have a "view." Fund letters must justify decisions. Traders themselves must preserve morale and client confidence. The simplest way to do that is to turn a squiggly line into a coherent plot: "Stocks fell because investors fear inflation," "The rally was driven by short covering," "The market is pricing in a pivot."

Sometimes those sentences are directionally true. Often they are post-hoc theater. The market is an aggregation of millions of motives across time horizons, constraints, and information sets. Compressing that into a single causal sentence is like watching a murmuration of starlings and concluding, confidently, that they turned left because one bird had a thought.

The Narrative Fallacy is our tendency to impose a neat story on messy reality, especially after we already know the ending. It feels like understanding. It's often just *compression*. A narrative is a zip file for complexity: it shrinks a chaotic world into something portable enough to carry around in the mind. The cost of that portability is accuracy.

The danger isn't merely that you'll believe a wrong story about yesterday. The danger is that stories change what you do *tomorrow*. A tidy narrative doesn't just explain; it persuades. It recruits you into a stance—bullish, bearish, outraged, smug—and once you have a stance, you start defending it. You selectively search for confirming evidence, you ignore disconfirming evidence, and you interpret ambiguity as support. Your brain stops doing Bayesian updating (adjusting beliefs based on new evidence) and starts doing public relations.

Narratives also have a social life. A story that spreads isn't necessarily the truest one; it's the most contagious—simple, emotional, identity-confirming, and easy to repeat. Traders like to think they are lone wolves. In practice, many are pack animals with Bloomberg terminals.

A useful way to think about narrative risk is to separate two questions that get conflated:

- Is the story *true*?
- Is the story *tradable* because other people will act as if it's true?

In the short run, the second question can dominate the first. In the long run, reality is still employed.

The market's favorite prank is to make a narrative tradable right up until you size it. Once you scale up, the costs that were invisible at small size appear like a bill slid across the table: slippage (worse execution than you expected), liquidity (not being able to exit without moving the price), and volatility (larger swings that hit your risk limits before your thesis has time to work). Many traders don't blow up because their story was wrong; they blow up because their *path* was wrong. They were right eventually and dead beforehand.

Hindsight Bias is the Narrative Fallacy's more dangerous sibling. After something happens, your mind rewrites your memory so the event feels predictable. The chart looks "obvious." The warning signs were "everywhere." The outcome becomes a moral referendum on everyone's intelligence. Then you look back at your own decision and judge it by the rewritten movie rather than the messy script you were actually holding at the time.

Hindsight Bias matters because it corrupts your data set. Your past trades are your laboratory. If the lab notebook keeps getting edited after the experiment, you can't learn cleanly. You start "remembering" that you had stronger conviction than you did, that you "knew" the catalyst was coming, that you "almost sold" before the crash. The ego loves these revisions because they preserve the feeling of competence. The trader pays for them in repeated mistakes.

There's an uncomfortable discipline that separates professionals from tourists: preserving the integrity of the *time-stamped self*. What did you believe then? What did you know then? What probabilities would you have assigned then? If you can't answer those questions without the glow of the outcome, you're not analyzing—you're mythmaking.

This is why serious traders write things down. Not because journaling is wholesome, but because it's adversarial. It's a way of trapping your future self in the truth. A good trade journal isn't a diary; it's a record of claims. It forces specificity: "If X happens, I'm wrong." "If volatility expands beyond Y, my sizing was too big." "If this gaps down, I will not add unless condition Z is met." Without that specificity, every outcome can be rationalized, and a rationalized outcome teaches nothing.

The market does not punish rationalizations immediately. It punishes them when you least feel like you're rationalizing—when you call it conviction.

Another mind trap hides inside the phrase "I was right." In markets, "right" can mean at least three different things:

- Right about *direction* (up or down).
- Right about *timing* (when).
- Right about *magnitude* (how much).

A narrative will often let you claim one "right" while ignoring the other two. "I was right about the fundamentals" can be a comforting sentence said by someone who went bankrupt being right at a 3-year horizon while trading a 3-week margin call schedule.

This is where probabilistic thinking stops being an intellectual badge and becomes a survival skill. A trade is not a declaration of truth; it's a bet with a distribution of outcomes. The correct question is not "Am I right?" It's "Given what I believe, what outcomes are plausible, and am I being paid enough for the risk I'm taking?"

If you want a mental model that keeps you honest, borrow one from poker: separate "playing well" from "winning the hand." Poker players live in the Bad Beat quadrant and don't collapse. They understand variance. They understand that the point of good play is not to win every hand, but to make decisions that win *over a large sample.* Markets are poker with worse lighting and more hidden cards.

The moment you accept that, a new kind of calm becomes available. Not the calm of certainty—markets rarely grant that—but the calm of coherence. You stop needing every tick to validate you. You stop interpreting a red day as an insult. You can take a loss without turning it into a personality crisis.

Of course, the brain doesn't give up its stories easily. It will keep trying to turn randomness into meaning because meaning feels safe. So the goal isn't to become a robot; it's to become bilingual: able to hear the story your mind is telling and simultaneously hold a more statistical, less theatrical interpretation.

A practical way to do that is to treat every seductive narrative as a hypothesis with an implied forecast. "The market is rallying because liquidity is coming back." Fine—what should we expect next if that's true? Lower credit spreads? Higher breadth? Specific sector leadership? If the story can't cash out into observable implications, it's not analysis; it's a vibe.

And when you do have a forecast, calibrate it. Most traders are not as wrong as they think; they're as *overconfident* as they don't realize. Overconfidence doesn't show up as saying "I can't be wrong." It shows up as assigning probabilities that are too tight, too soon. The trader who says "This is a 90% setup" is often not claiming omniscience; they're leaking emotion into math.

Calibration is the antidote: learning what your "70% confidence" actually means in reality. If you label fifty trades as 70% likely to work and only twenty-five work, you're not unlucky—you're miscalibrated. The market didn't break; your probability language did.

There's also a darker narrative trap: moralization. Traders love to turn markets into a justice system. Shorts are "evil." HFT is "rigged." The Fed is "saving" or "destroying" everyone. These stories can contain pieces of truth about incentives and structure, but they often function as emotional outsourcing. If the market is an antagonist, then your losses can be blamed on villainy rather than on sizing, process, and uncertainty.

The reality is colder and more useful: markets are not fair, but they are mostly indifferent. Indifference is great news. You don't need to win an argument with the universe. You need to place bets that make sense under uncertainty and survive long enough for the edge—if you have one—to express itself.

The final trap in this cluster is identity fusion: letting a position become "me." The moment a trade becomes identity, evidence becomes offensive. A neutral data point ("earnings were weaker than expected") turns into a personal attack ("they don't get it," "the market is stupid," "this is manipulated"). At that point, you're no longer managing risk; you're defending a self-concept.

A trader who can't say "I'm wrong" is a trader who can't exit. And a trader who can't exit eventually meets the one trade that refuses to negotiate.

A clean way to keep identity separate is to frame positions as *provisional*. Not weakly held—provisional. Held with conditions. Held with respect for alternative explanations. Held with an explicit acknowledgement that the market may reveal information you don't yet have access to. That's not humility as a virtue. That's humility as a tactic.

It helps to remember what a market price is: it's not a verdict on your intelligence. It's a clearing price where urgent buyers and sellers meet, shaped by constraints you don't see (risk limits, hedging flows, redemptions, margin, index rebalances), by time horizons you don't share, and by information you may not have yet. Prices are sometimes wrong about fundamentals. They are rarely wrong about *positioning risk*—the risk that too many people are leaning the same way and the exit is narrower than the entrance.

So the mind needs guardrails:

- Don't grade decisions by single outcomes.
- Don't accept narratives that can't be translated into forecasts.
- Don't let hindsight rewrite the evidence you learned from.
- Don't let identity buy stock on margin.

None of this eliminates emotion. It makes emotion less powerful as a trading signal.

A good trader isn't the person who feels nothing. It's the person who can feel the full punch of a loss, notice the mind reaching for a story to anesthetize it, and still return to the only question that matters: *Was the bet well-constructed given what I could reasonably know at the time, and did I size it so I could keep playing?*

That's decision quality. It's quieter than the P&L. It's also the only thing that compounds.

3.2 The Wisdom and Madness of Crowds: Incentives and Debiasing

A crowd is not merely a lot of people. It's a lot of *consequences*. The moment you put money, reputation, career trajectory, and dinner-table identity into the same room, "belief" stops being a quiet private thing and becomes a social instrument. In markets, that instrument is tuned to incentives. Not lofty incentives like "maximize truth", but practical ones like "don't get fired", "don't look stupid", and "don't underperform the neighbor you secretly hate."

That's why the most dangerous market mistakes aren't always analytical. Plenty of disasters start with a correct observation and end with a distorted action. The distortion comes from social gravity—everyone's awareness of everyone else's awareness—plus the uncomfortable fact that professional investing is often scored on relative performance over short windows, even when the underlying game is long and uncertain.

Consider the fund manager who thinks a stock is wildly overvalued. If she underweights it and it keeps ripping, she doesn't just lose money; she loses *permission*. Clients question her competence. Consultants remove her from recommended lists. Her own firm starts watching the monthly numbers like hawks watching a field mouse. Even if she's right in the end, she may not be around to collect the payoff. That's career risk: the risk that being early looks identical to being wrong.

So she buys it anyway—perhaps a little, perhaps a lot—because the position is no longer about valuation. It's about avoiding isolation. When everyone owns the same hot thing, the trade becomes socially insured. If it blows up, you weren't uniquely foolish; you were merely human in a very crowded room. Nobody gets celebrated for originality in a bubble, but plenty of people get punished for refusing to participate.

Retail traders don't have consultants and redemption cycles, but they have their own invisible committee: group chats, online communities, the curated highlight reels of strangers. Retail incentive systems are emotional rather than institutional, and they can be more intense. A person who publicly identifies as "a crypto guy" or "an EV believer" is no longer simply holding an asset. He's defending a badge.

That badge creates a strange asymmetry: changing your mind feels like losing status, even when changing your mind is exactly what good decision-making looks like. The trade becomes a referendum on self-worth. And once the trade is self-worth, selling at a loss isn't "closing a position." It's confession.

Markets are filled with these private incentive contracts. People don't say them out loud, but they trade as if they signed them. "I must be right." "I must not look weak." "I must not miss this." These are not trading rules; they're social rules wearing a Bloomberg costume.

Herding—copying what others do—often gets described as stupidity. That's comforting, because it implies you can avoid the crowd by being smart. But herding is frequently rational at the individual level. When you don't fully trust your own information, the actions of others become data. If everyone is buying, maybe they know something. Even if they don't, maybe the price will go up anyway because more people will copy them. At that point the herd isn't just a herd; it's a self-propelling mechanism.

This is how informational cascades form: once enough people act in one direction, later participants can find it optimal to ignore their own private signals and follow the observed crowd. A cascade can start from something small—an early piece of news, a charismatic voice, a few big prints in the tape—and then it hardens into "what the market thinks." The logic is brutal: you might privately suspect the crowd is wrong, but you also suspect the crowd can stay wrong longer than you can stay solvent.

Cascades have a tell. Early on, the story is "Here are the facts." Later, the story becomes "Look at the price." Price is a form of evidence, but it's also a form of *marketing*. When a narrative is strengthened by price action, confidence rises faster than information quality. People mistake momentum for proof. Conviction becomes contagious. At the extreme, disbelief becomes a personality flaw: "You just don't get it."

The crowd can be wise. Aggregation works when participants bring diverse information and incentives are aligned with truth-seeking. But crowds become mad when everyone watches everyone else, when reputational payoffs dominate analytical payoffs, and when volatility turns patience into a luxury.

A market ecology built on incentives creates predictable distortions. Not because people are evil, but because they're employed. Sell-side research is supposed to inform, yet it's entangled with relationships. Corporate executives speak to investors with one eye on the stock price. Brokers route orders through systems shaped by rebates, internalization, and execution quality tradeoffs. Even the most conscientious professionals operate inside a web of constraints: compliance, mandates, tracking error limits, risk models, benchmark pressures, client optics.

This doesn't mean "markets are rigged." That phrase is a narrative that offers emotional relief at the price of useful detail. A better sentence is: *markets are full of agents optimizing different objective functions*. When you understand the objective functions, behavior stops looking random. It starts looking structural.

The practical question is not "How do I avoid all bias?" but "How do I keep my decision quality from getting hacked by incentives—mine and everyone else's?" The answer is decision hygiene: a set of habits that protect the thinking process the way good sanitation protects a kitchen. You're not trying to eliminate all germs. You're trying to stop preventable infections from ruining the meal.

The first principle of decision hygiene is separation: separate identity from position, separate analysis from execution, separate thesis from price action, separate risk management from mood. Most trading mistakes happen when these collapse into one sticky emotional blob.

A concrete example: imagine you're bullish a stock because your work suggests margins will expand. Then the stock drops 6% in a day on no fundamental news. If your identity is fused with "I'm right," the drop feels like disrespect and you may add impulsively to prove a point. If your analysis is fused with execution, you may interpret the drop as new information even if nothing changed. If your risk management is fused with mood, you may widen stops or remove them entirely because you can't bear realizing pain.

Decision hygiene is what prevents that chain reaction.

A surprisingly effective technique is *blind analysis*: build the thesis without seeing the price chart or knowing the ticker. Analysts hate this idea at first because it feels like giving up an advantage. In reality, it removes a contaminant. Price is a powerful priming device. If you know something has doubled, you subconsciously look for reasons it should keep doubling. If you know it has collapsed, you subconsciously look for reasons it's broken. Blind analysis forces you to evaluate the business, the balance sheet, the competitive position, and the macro exposure without your brain already holding a verdict.

The process can be as strict or as light as you want. At the strict end, one person prepares a "company packet" that redacts the name, ticker, and chart, leaving only fundamentals and key filings. At the light end, you simply delay looking at price until you've written down your thesis and the disconfirming evidence that would kill it. Either way, the point is the same: stop letting the crowd's opinion (embedded in price) pre-write your conclusion.

Blind analysis also protects against one of the market's quieter vanities: the desire to be seen as someone who "called it." People love to buy things that are already rising because the chart gives them a story to tell. "I'm in this." "I've been on this." The chart becomes a social accessory. Blind analysis forces you to purchase your confidence with reasoning rather than with social proof.

The next technique is the *pre-mortem*. It sounds theatrical, but it's basically a structured way to trick your brain into honesty. Instead of asking "What could go wrong?", you assume the trade *did* go wrong and ask: "It's six months from now. This position is a painful loser. What happened?"

That small grammatical shift matters. "Could go wrong" invites optimism and vagueness. "Did go wrong" invites specificity. It gives your mind permission to generate failure scenarios without feeling disloyal to the idea. It turns worry from an emotion into an investigation.

A good pre-mortem produces a list that is uncomfortable and therefore valuable: the competitor you're underestimating, the regulatory risk you waved away, the cyclicality you denied, the liquidity you assumed would be there on the exit. It's also where social incentives show up. "Everyone piled into this after the podcast episode." "My conviction increased because the stock went up." "I ignored the short report because it threatened my identity." Those are not fundamental risks. They are *process risks*. And process risks are the ones that keep repeating.

Pre-mortems work best when you force them into numbers. Not because numbers are magical, but because they are less emotionally pliable. If a pre-mortem scenario is "Margins disappoint," ask how much. If it's "multiple compresses," ask from what to what. If it's "liquidity vanishes," ask what volume looks like and what your exit would realistically cost. Vague fears are easy to ignore; quantified risks demand a sizing response.

The third hygiene tool is the one that sounds boring and is therefore underused: *decision checklists*. Aviation uses checklists not because pilots are dumb, but because pilots are human under stress. Trading is also human under stress, with one extra problem: the stress can be self-generated. You can have stress on a perfectly quiet day simply because your P&L is loud.

A trading checklist is not a bureaucratic hoop. It's a pre-commitment device. It helps you avoid the kind of mistake that makes you stare at the ceiling at 3 a.m. saying, "I can't believe I did that." The best checklists are short and personal: the handful of errors you are most likely to repeat because they feel good in the moment. Overtrading. Revenge trading. Adding to a loser without new information. Taking profits early because you want relief. Turning a trade into an investment because selling would mean admitting defeat.

The checklist is where you install your own psychological guardrails.

One of the most powerful checklist items is also one of the simplest: *What is my edge here?* "Edge" means you have some reason to believe your expected value is positive—your odds and payoff are favorable compared to the price. If you can't articulate the edge in plain language, you're probably renting the crowd's confidence.

Another is: *What is the non-consensus part of this?* If your thesis is identical to what everyone already believes, your potential upside is often just "things continue as expected." That can still be a trade, but it should change sizing and time horizon because surprises tend to be where the money is made—and where it's lost. Crowded trades are not automatically bad. They are just fragile. They require humility about exits.

A third is: *What would I need to see to reduce or exit?* Most people can list reasons to add. Fewer can list reasons to leave. Leaving feels like abandonment; adding feels like courage. That's an emotional bias dressed as "conviction." A pre-defined exit condition is not a lack of belief; it's respect for uncertainty.

Debiasing isn't only about individual technique. It's also about designing your environment so your incentives don't quietly poison your thinking.

If you're a professional investor, one of the cleanest environmental fixes is to separate the act of forecasting from the act of risk-taking. Some firms do this formally: analysts propose, portfolio managers size. Even if you do both roles yourself, you can simulate the separation by creating a time delay. Write the thesis and probability distribution today; size tomorrow. Time is a solvent. It dissolves the urgency that crowds create.

Another environmental fix is to stop consuming market commentary as entertainment. Entertainment is not neutral. It's engineered to provoke emotion—fear, outrage, FOMO, moral certainty—because emotion holds attention. If your information diet is designed to spike your nervous system, don't be surprised when your trades look like nervous-system artifacts.

This is not a call to be uninformed. It's a call to be deliberate. Choose a small set of high-signal sources. Build a habit of reading primary material when it matters: filings, transcripts, data releases. Even a modest shift toward primary sources reduces narrative infection because you're not ingesting someone else's story about the story.

Incentives also distort through time horizon. A retail trader with a day job may think he's "long-term," but he checks his position every ten minutes. A professional fund may claim to be a three-year investor, but clients can redeem monthly. The mismatch between stated horizon and lived horizon creates a constant temptation to manage optics. Optics are the enemy of expected value.

One of the simplest ways to reduce time-horizon distortion is to predefine the evaluation window for a trade. Not the maximum holding period—the *evaluation* period. If your thesis depends on an earnings cycle, evaluating yourself after two trading days is pure resulting. It's like pulling a cake out of the oven after five minutes and declaring the recipe bad.

A disciplined trader decides in advance what "enough time" means for information to arrive. Then, within that window, price action is treated as data about positioning and volatility—not as a verdict on the thesis. This single habit lowers the volume of the crowd in your head.

Crowds also create a subtler trap: the illusion of consensus as safety. When everyone believes something, it feels less risky. In reality, consensus can be the source of risk. If everyone is on one side, the marginal buyer disappears. The trade becomes dependent on continuous recruitment: more believers, more flows, more narrative oxygen. When the oxygen thins, prices can gap because the exit is crowded.

That's why "everyone owns it" is not a comforting fact. It's a structural fact about liquidity. When you internalize that, you stop asking only "Is this true?" and start asking "How crowded is this, and what happens if the crowd needs the door at the same time?"

Decision hygiene for crowd risk often looks like boring portfolio craft: smaller size in crowded themes, wider respect for volatility regimes, and a willingness to take partial profits when the narrative gets too perfect. Not because you've "lost faith," but because narratives at peak confidence tend to underprice surprise.

The punchline is not cynicism. The goal isn't to sneer at crowds or pretend you're above social influence. The goal is to recognize that the market is a social organism built on incentives, and then to build a private decision system that doesn't get dragged around by the organism's moods.

The market will always offer you a trade in exchange for your identity: "Join the consensus and you'll feel smart." It will also offer you the opposite trade: "Refuse the consensus and you'll feel superior." Both are traps, because both make the trade about you.

A clean process is quieter. It asks: What do I believe? How uncertain am I? What would change my mind? How large can I be and still think clearly? The crowd can be a source of information, but it is never a substitute for a decision.

And when the crowd is loud—when narratives are viral, screens are flashing, and everyone seems to be making easy money—decision hygiene becomes not just a best practice, but a kind of moral courage: the courage to be unentertained, to be unhurried, and to let the market's social weather pass through without rearranging the furniture in your brain.

3.3 The Playing Field: Market Ecology and Trading Realities

Most investors talk about "the market" the way people talk about "the weather": a single thing that happens to them. But price is not weather. Price is a negotiation. And like any negotiation, it depends on who shows up, who's desperate, who's patient, who has rules, and who is allowed to walk away.

That's market ecology: the study of the different *species* in the financial habitat-retail traders, institutions, market makers, central banks-and the flows between them. Not flows of opinions. Flows of *liquidity*: the ability to buy or sell without moving the price too much. When liquidity is abundant, markets feel calm and fair. When it disappears, the market stops feeling like a spreadsheet and starts feeling like a crowd trying to squeeze through a narrow door.

Liquidity is easy to romanticize as "depth," but in practice it's a set of promises that can be withdrawn. The bids you see are not vows; they're invitations with fine print. That fine print is where trading realities live: routing incentives, order types, volatility controls, and leverage mechanics that turn a small mistake into a forced liquidation at the worst possible moment.

The first step toward decision quality is accepting an unflattering truth: you are not trading against "the market." You are trading against other participants' constraints. Sometimes those constraints make them predictable. Sometimes they make them dangerous. Always they matter more than your narrative about what "should" happen.

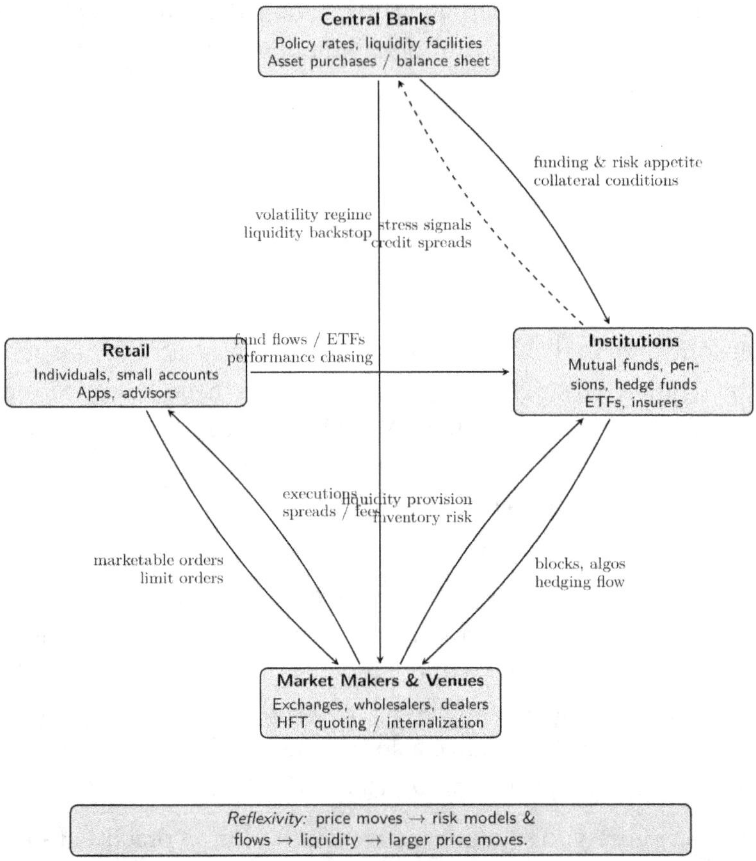

Once you see the ecosystem, a lot of "mysteries" stop being mysterious. A midday air pocket, a violent close, a rally that ignores bad news-these often come from who was forced to trade, not from who had the best argument.

Start with retail. Retail is not small in aggregate, but it tends to be fragmented: lots of accounts, lots of short-horizon decisions, and a strong preference for simplicity. Retail flow is often *marketable-*

meaning it prioritizes getting a fill now. That preference matters because "now" is expensive in volatile moments. The market happily sells immediacy at a markup, and the markup widens when everyone wants immediacy at the same time.

Retail also brings a peculiar kind of liquidity: liquidity that can vanish because of emotion rather than mandate. Panic selling isn't a strategy; it's a loss of time horizon. In quiet markets, that doesn't stand out. In fast markets, it becomes fuel.

Institutions are different animals. A pension fund may be slow and rule-bound. A macro hedge fund may be fast and opportunistic. An ETF may be mechanically forced to transact when money flows in or out. The defining feature is not intelligence; it's constraint. Institutions have mandates (what they are allowed to own), risk models (how much they are allowed to lose), and clients (who can withdraw capital at awkward times). Those constraints create patterns: rebalancing near month-end, de-risking when volatility rises, forced selling when redemptions hit.

The most underestimated institutional force is the "risk manager inside the machine." Many funds don't decide to sell because their thesis changed; they sell because their *volatility* changed. Volatility is not merely drama on a chart. It's a hard input into leverage limits, margin requirements, and value-at-risk style controls. When volatility spikes, the same position suddenly becomes "bigger," even if you didn't add a share. That can trigger automatic or semi-automatic de-risking. The selling then increases volatility, which triggers more selling. Ecology turns into weather.

Market makers sit at the center like the circulatory system. Their job is to quote two-sided prices-bid and ask-and earn the spread while managing inventory risk. In plain language: they try to buy from sellers and sell to buyers without getting stuck holding too much of a falling asset or being short too much of a rising one.

Market makers are often described as if they are altruistic liquidity fountains. They're not. They are liquidity *renters*. They provide liquidity when they can hedge and when risk feels measurable. They step back when they can't hedge, when the news is discontinuous, or when volatility makes every quote a potential loss. This is why liquidity disappears exactly when you most want it. The very conditions that make you desperate are the conditions that make liquidity providers cautious.

And then there are central banks, which do not "trade" like the other species but still shape the ecosystem the way gravity shapes a river. Policy rates influence the cost of leverage and the attractiveness of cash. Liquidity facilities influence whether financial plumbing clogs during stress. Asset purchases and balance sheet policy can affect term premiums, risk appetite, and the market's sense of what "normal" volatility should be. Central banks don't decide whether your next trade works. They do influence the regime in which your trade lives: easy funding and tight spreads versus expensive leverage and jumpy markets.

Now layer in structure: in U.S. equities, you are not trading on "the exchange." You are trading in a fragmented landscape of exchanges, wholesalers, and alternative venues, stitched together by regulation and technology. One key piece of that stitching is the idea that brokers should not execute you at a price worse than protected quotations elsewhere (the intuition behind the Order Protection Rule). That sounds like a simple promise: best price wins.

But "best price" is not the full story because execution is a multidimensional product. A penny better quote isn't helpful if you can't access it in size, or if you get partial fills that leave you exposed, or if latency and queue position mean you're always last in line. Brokers also make routing decisions-where your order is sent-that can be influenced by economics such as rebates or payment for order flow arrangements. Those incentives don't automatically imply misconduct, but they do mean you should stop imagining your order as a sacred message delivered to a neutral god. Your order is a piece of commerce moving through a system with its own business models.

This is where "free trading" earns its quotation marks. You may pay zero explicit commission and still pay in spreads, in opportunity cost, or in execution quality. The cost is just less visible, which makes it easier to ignore—right up until you try to trade a fast move and discover that your fill price is an opinion the market did not share.

Order types are the retail trader's most immediate lever for navigating this ecology, and they are regularly misunderstood because platforms make them look like buttons rather than contracts.

A *market order* means: "Fill me now, whatever the price." It buys certainty of execution by selling away control of price. In quiet markets, that tradeoff can be reasonable. In a volatility spike, it can be brutal. If liquidity thins and quotes gap, the market order obediently follows the gap.

A *limit order* means: "Fill me at this price or better." It buys control of price by accepting uncertainty of execution. The hidden cost of limit orders is not just missed fills; it's *selection*. You are most likely to get filled when the market is moving against you-when your limit becomes attractive to the other side. That doesn't make limit orders bad. It makes them honest: they force you to confront the difference between "I want this price" and "the market will give me this price."

Stop orders are the most emotionally appealing and the most treacherous in fast markets. They feel like safety—"If it hits this level, get me out"—but many stops become market orders when triggered. In a gap, that can mean your "get me out at 100" turns into "sell me somewhere in the neighborhood of whatever is available," which is how traders learn that risk is not a number; it's a path.

All of this gets amplified by leverage. Margin trading is not merely "buying more." It's inviting the broker into your decision-making, because margin introduces the possibility of forced liquidation. If the position moves against you far enough, the broker can require more collateral. If you can't provide it in time, positions may be sold to meet requirements. That can happen at precisely the moment liquidity is worst and spreads are widest. Leverage turns volatility into a solvency question.

This is why two traders can make the same trade idea and have opposite outcomes. One sized it so the market had time to be noisy. The other sized it so noise became fatal. The ecology doesn't care which story you believed; it cares whether your position can survive the environment.

Volatility itself is not merely "more movement." It changes behavior. It changes quoting. It changes which strategies function. It changes the *type* of liquidity you can access. In calm markets, liquidity feels continuous: you can scale in and out and the tape is forgiving. In stressed markets, liquidity becomes discrete: pockets of emptiness, sudden gaps, and price levels that trade like trapdoors.

A useful mental reframe: liquidity is like oxygen, and spreads are like the price of breathing. When oxygen is plentiful, you don't think about it. When it's scarce, every breath becomes expensive, and people make bad decisions just to stop the discomfort. Traders do the same thing: they cross wide spreads to relieve anxiety, they smash market orders to regain a sense of control, they "simplify" complex risk by exiting at the worst time.

Then there's reflexivity—the loop where prices don't merely reflect fundamentals but also *change* them. Higher stock prices can lower a company's cost of capital, enabling investment, acquisitions, and employee retention through equity compensation. They can also attract passive flows and momentum strategies, increasing demand simply because the price went up. The reverse is also true: falling prices can tighten financing, trigger covenant pressure, spook customers and suppliers, and force dilution. In reflexive regimes, the market is not a mirror. It's a steering wheel.

Reflexivity is why "the market is irrational" is often an incomplete complaint. The market may be expressing second-order realities: forced flows, collateral constraints, risk-parity deleveraging, volatility-targeting adjustments, ETF creations/redemptions. These are not fairy tales. They're mechanisms. When you can name mechanisms, you can trade with less superstition.

Market ecology also changes the meaning of common trading advice. "Buy the dip" is not a strategy; it's a bet on who is forced to sell and whether liquidity providers are still standing there catching the flow. "Follow the trend" is not a strategy; it's a bet that the marginal buyer has not been exhausted and that the exit will remain wide enough when you need it. The ecology determines whether those slogans are profitable heuristics or expensive memes.

So what does a retail trader do with this without drowning in microstructure detail?

First, stop treating execution as an afterthought. In an ecology where liquidity can evaporate, "good idea" and "good trade" diverge. Use order types intentionally, not habitually. When volatility is high, be suspicious of market orders. When you place limits, understand you're negotiating with the other side, not requesting a favor.

Second, treat leverage as an environmental multiplier. If you use margin, your true risk is not your stop-loss level. Your true risk is the combination of price movement, volatility expansion, and liquidity conditions that can produce forced actions. The market's worst days are not polite enough to wait for your plan.

Third, learn to recognize which species is likely driving the tape. A slow drift can be institutional accumulation. A violent spike into the close can be hedging and rebalancing. A sudden air pocket can be liquidity providers widening out. You will not be right every time. But even a rough read is better than trading as if every move is a referendum on your intelligence.

Finally, accept that the market is not one opponent; it's a table full of opponents playing different games. Some have faster information. Some have cheaper execution. Some have mandates that force them to do dumb things at predictable times. Your edge-if you have one-often comes from choosing the games where those structural forces are favorable to you.

That's the playing field. Not a philosophy seminar, not a morality play: an ecosystem of incentives and constraints, where liquidity is the food, volatility is the climate, and survival is the prerequisite for learning anything at all.

Chapter 4

Finding an Edge: Evidence, Data, and Research

If investing were merely about having good ideas, we would all be billionaires. The graveyard of trading accounts is filled with 'sure things' that worked perfectly in theory but crumbled in reality. Finding a true edge isn't just about spotting a pattern; it's about acting like a scientist rather than a gambler. In this chapter, we move from gut instinct to evidence. We will explore how to interrogate data until it tells the truth, how to avoid the optical illusions that fool our brains, and how to distinguish between a genuine market signal and a lucky streak.

4.1 From Idea to Edge: Building Testable Hypotheses

A trading idea usually shows up the way a smell does: suddenly, and with unreasonable confidence. You notice tech acting well, or small caps "feeling heavy," or oil rallying even on bearish headlines. The mind loves fast stories. It stitches together a handful of observations, adds a pinch of recent profit or pain, and calls it a thesis. The problem isn't that intuition is useless. The problem is that intuition is *not a claim*. It's a mood. And markets do not pay you for moods.

The moment you decide to risk money, the idea has to survive a harsher environment than your imagination: it has to survive contact with time, costs, and randomness. That means converting your hunch into something that can be wrong in a clear, unambiguous way. Not "I think semis are strong," but "When X happens, Y tends to follow within Z days, often enough and large enough to beat costs." Notice how that sentence is already less fun. Good. Fun ideas are easy to invent. Investable ideas are hard to *corner*.

A useful way to start is to treat a trade like a courtroom case. Your hunch is the opening statement. A testable hypothesis is admissible evidence. "Tech looks strong" is not admissible because it doesn't specify what "tech" means, what "strong" means, compared to what, over what time window, with what risk, and under what conditions you'd admit you were wrong. If you can't write down the conditions under which your belief changes, you don't have a belief. You have a preference.

The gap between a preference and a hypothesis is where most of the damage happens. Traders blow up not because they never had good ideas, but because their ideas were never forced into a form that could be audited. They were allowed to shapeshift. When the trade goes against them, the definition of "the setup" quietly changes. When it goes in their favor, the story hardens into destiny. This is outcome worship in disguise: judging the quality of your thinking by whether you got paid. The discipline here is the opposite. You judge your thinking by whether you *specified reality in advance.*

A hypothesis has three parts: a mechanism, a measurement, and a set of disconfirming outcomes. The mechanism is the *why*. The measurement is the *what exactly*—the data you will compute with no wiggle room. The disconfirming outcomes are the *how you'll know it failed.* Without a mechanism you risk fitting patterns that are accidental. Without measurement you can't test. Without disconfirmation you can't learn, because you've built a story that can explain everything, which is another way of saying it explains nothing.

Mechanism first, even if it's messy. Markets are not random-number generators; they are crowds of constrained decision-makers. When prices move, something usually changed: information, liquidity, risk tolerance, positioning, regulation, or the cost of leverage. A mechanism is your proposal for which of those gears is turning. "Rates down means growth up" is a mechanism, but it's so wide it's basically a weather forecast. Tighten it: *which* rates, *why* growth benefits, *what* channel dominates (discount rate, earnings, risk appetite), and *when* it should matter (immediately, or with lag). The more specific your mechanism, the more your hypothesis can be wrong in a way that teaches you something.

Then measurement: translate the mechanism into variables you can compute as if you were a stranger trying to reproduce your work. If the mechanism is "institutional flows chase recent winners," measurement might be "stocks in the top decile of 12-month returns, skipping the most recent month, outperform over the next month." If the mechanism is "volatility sellers get squeezed after calm periods," measurement might be "after 20 trading days of unusually low realized volatility, the next volatility spike is larger than average." You are turning a story about human behavior into a rule that a computer could follow without knowing your backstory.

The third part—disconfirmation—is where the ego usually starts negotiating. Most people write hypotheses like they write horoscopes: in a way that protects the author. "Tech should outperform unless there's macro uncertainty or a rotation into defensives." That sentence is a shield. It is designed to keep the idea alive no matter what happens. Real disconfirmation is sharp. It has teeth. It looks like: "If the edge exists, average next-month return should exceed the benchmark by at least 30 basis points after estimated costs, and the effect should persist across at least three non-overlapping market regimes; if the effect disappears in out-of-sample data, we treat it as noise." That's not pleasant. It is, however, honest.

The trick is to define what you mean by "edge" before you go hunting for it. An edge is not "a pattern I like." An edge is a repeatable advantage that survives friction: spreads, fees, financing, slippage, taxes, and the fact that your execution won't be perfect. It also survives your own behavior: missing signals, sizing too large after a win, freezing after a drawdown. A hypothesis that ignores frictions is a hypothesis about a fantasy trader operating in a frictionless universe. That trader is undefeated. You are not.

Most hypotheses also fail because the idea is under-specified. Here's what under-specified sounds like in the wild:

> "Buy quality stocks in a rising rate environment."

It's a vibe, not a model. "Quality" can mean margins, stability, low leverage, high ROE, low accruals, high governance scores, or whatever factor a backtest needs it to mean. "Rising rates" can be nominal, real, short-end, long-end, expected, or surprise. The "environment" can be two weeks or two years. With enough flexibility, you can *always* find a version that looks good in hindsight. That's not research; it's costume design.

A good hypothesis removes degrees of freedom. Degrees of freedom are the hidden dials you can twist until the past looks flattering: which universe, which start date, which rebalance frequency, which filter, which benchmark, which definition, which lag, which cost assumption. Every dial is a chance to accidentally (or not-so-accidentally) fit noise. The goal is not to eliminate judgment—judgment is the whole game—but to quarantine it. Decide the dials up front. Then test.

Here's a practical template that forces clarity without turning you into a statistician:

Component	*Write it like this*
Universe	"Trade U.S. listed common stocks above $2B market cap, excluding REITs and ADRs."
Signal	"At month-end, compute 12-month total return skipping the most recent month; rank into deciles."
Rule	"Go long top decile, short bottom decile; equal-weight; rebalance monthly."
Holding period	"Hold for one month; close and refresh at next month-end."
Risk control	"Target 10% annualized volatility; sector-neutral within GICS sectors."
Friction model	"Deduct 15 bps per side; include borrow cost estimate for shorts."
Prediction	"Expected monthly alpha $>$ 20 bps net; hit rate $>$ 52%."
Disconfirmation	"If net alpha $<$ 0 over the next 5 years out-of-sample or concentrated in one regime, reject."

The purpose of writing this down isn't to sound rigorous at a dinner party. It's to stop your future self from lying to you. Your future self is the most dangerous person in your trading career because they will have a compelling narrative and selective memory. A written hypothesis is a contract with that future liar.

You'll notice the template forces you to make an awkward choice: are you trying to predict returns, or are you trying to build a portfolio that behaves a certain way? Many investors confuse these. They say they have an "edge" when they really have a *tilt*. A tilt is a persistent exposure to a risk factor (value, momentum, carry, quality) that has some long-run expectation but can underperform for years. An edge is something closer to a repeatable mispricing or structural advantage—often smaller, more fragile, and more dependent on execution. Both can be worth owning, but they must be tested and managed differently. A tilt can be a long-term policy. An edge is a job.

Mental models are the bridge between mechanism and measurement. They are simplified stories about how the market behaves that help you predict what matters. The best mental models are not grand theories; they're small machines. They take inputs and produce outputs. Here are a few that regularly show up in real trading, stated as machines rather than slogans:

Crowded trade model: When many participants hold the same position, bad news causes forced selling that amplifies moves. Inputs: positioning proxies, funding conditions, volatility. Output: fat left-tail risk and sudden correlation spikes.

Liquidity vacuum model: When natural buyers step away (risk-off, regulatory constraints, dealer balance sheet limits), prices must move further to find marginal demand. Inputs: spreads, depth, volatility regime. Output: gap risk and slippage.

Narrative adoption model: Prices don't move only on information; they move when a story becomes socially tradable. Inputs: surprise, media attention, analyst revisions. Output: trend persistence until the narrative saturates.

You don't need to believe any of these are always true. You need them to be specific enough that they produce predictions *you can embarrass*. "Crowding causes crashes" is not embarrassing. "When my crowding metric is in the top decile and funding spreads widen, downside tails in the next month widen by X" can be embarrassed. That's progress.

The most honest question you can ask at the idea stage is: *What must be true for this trade to work?* Not "what do I hope happens," but what structural assumptions are doing the heavy lifting. If your thesis is "AI capex is a multi-year cycle," what must be true?

- End customers must keep demanding AI-driven products at prices that justify spending.
- Companies must keep access to cheap enough capital to fund capex.

- Supply chains must scale without destroying margins.
- Competition must not erase profits faster than growth arrives.

Each "must be true" can be tied to observable proxies: order books, margins, credit spreads, capacity utilization, pricing. Now your thesis has handles.

Handles matter because markets punish vagueness with optionality. If your hypothesis is vague, you can reinterpret any outcome as "sort of consistent." You'll keep placing the same bet, but with shifting rationale. That's not adaptation; it's denial dressed up as flexibility. The goal is the opposite: create a hypothesis so crisp that reality either supports it or humiliates it quickly.

A good hypothesis also declares its time horizon like it's a legal identity. "This is a 5-day effect" and "this is a 5-year effect" are different species. Short-horizon edges tend to be more about microstructure, liquidity, and positioning. Long-horizon edges tend to be about valuation, structural growth, and macro cycles. Mixing them creates self-inflicted confusion: you buy a "long-term" idea and then stop out because it didn't work in a week; or you run a short-term mean-reversion strategy and then rationalize a drawdown as "the thesis takes time." Time horizon is not a detail. It's the whole bet.

Another underappreciated point: your hypothesis should specify the *payoff distribution*, not just the average return. Two strategies can have the same backtested Sharpe ratio and feel totally different in the body. One makes small gains most days and occasionally gets hit by a truck. The other bleeds slowly and occasionally jumps. The average can be identical while the experience is opposite. When you write your prediction, include at least one statement about tails: frequency and size of drawdowns, and what market conditions tend to produce them. That's not pessimism; that's risk management with eyes open.

Many investors fall in love with hypotheses that are actually *identity tricks*. Example: "High dividend stocks outperform." If the high dividend universe tilts toward value and low volatility, you may simply be rediscovering known exposures. That's not necessarily bad, but it's not new information. The right hypothesis would clarify whether the expected return comes from a known factor premium, from a behavioral mistake, or from a structural constraint (like mandates). If you can't say why the market should pay you, you may be confusing a label for an engine.

Now for an uncomfortable truth: markets are a pattern factory. If you stare at enough data, you will find "signals" that look like genius. This is not because you're dishonest; it's because the human brain is designed to detect structure, and financial data is noisy enough to offer endless false structure. The cure is not cynicism. The cure is a process that assumes your brain is a talented con artist and requires receipts.

One of the best receipts is a pre-committed prediction. Before you run the backtest, write down what you think you'll see and why. Not in vague terms, but in sign and shape: "Effect strongest in small caps," "decays over 3 months," "breaks during funding stress," "sensitive to transaction costs." Then test whether the data agrees. You're not just testing the strategy; you're testing whether you understood the mechanism. When the data contradicts your prediction, you have learned something. When the data matches, you still haven't proven truth—but you've earned the right to continue.

The second receipt is replication across slices that make economic sense. If your hypothesis is about liquidity, it should be stronger where liquidity is scarce: small caps, after-hours, stressed markets. If it's about attention, it should show up around events, news, and analyst coverage changes. Slicing the data is dangerous when done after the fact to rescue a weak result, but it is powerful when done as a *prediction from the mechanism*. That distinction is everything.

The third receipt is to make your hypothesis operationally realistic. Ask, with zero romance: can I actually trade this? If your signal requires buying the exact close, it might be untradeable. If it requires shorting hard-to-borrow names, your returns might be a borrow-cost illusion. If it turns over 400% a month, friction will eat it. If it concentrates in a few names, your capacity is limited. Hypotheses that ignore implementation often look incredible right up until money touches them. Then the edge evaporates like a magician's coin.

Even for discretionary traders, a hypothesis can be explicit without being robotic. "If high-yield spreads widen by X and the equity index is down Y over Z days, I will reduce gross exposure by N% and avoid adding leverage for two weeks." That's a hypothesis about your own behavior interacting with market regimes. It can be tested. Many traders would improve faster by testing *their decision rules* than by hunting for ever-more exotic signals.

One more psychological trap: the urge to make the hypothesis *beautiful*. Investors love elegant narratives because elegance feels like truth. Markets are not obligated to be elegant. Sometimes the best hypothesis is weirdly narrow: "This effect shows up only in the first three trading days of the month," or "only after earnings when guidance is raised but the stock gaps down." Narrow hypotheses can be legitimate if they map to a real constraint (index rebalancing, payroll flows, options hedging). The standard for narrow ideas is not that they sound deep; it's that they survive outside the sample and are tied to a plausible mechanism. Beauty is optional. Survival isn't.

The final step is to write the hypothesis in a single sentence that a skeptic would respect. Not because it's persuasive, but because it's falsifiable. Here are a few examples of what "skeptic-respectable" sounds like:

> "If U.S. large-cap equities close above their 200-day moving average and credit spreads tighten over the same month, then trend-following exposure produces higher risk-adjusted returns over the next 3 months than in months where spreads widen, net of conservative trading costs."

> "When implied volatility is in the bottom quartile of its 1-year range and dealer gamma positioning is estimated to be strongly short, the probability of a 2% single-day move in the index over the next 10 trading days is higher than baseline."

> "Within a defined universe of liquid futures, a carry signal (roll yield) predicts next-month excess returns after controlling for momentum, and the effect weakens during funding stress as proxied by rising short-term funding spreads."

Each sentence declares the inputs, the condition, the output, the horizon, and at least one place where it should fail. Each can be tested without mind-reading the author.

That's the transformation: hunch to hypothesis, story to specification, confidence to contracts with reality. The market is not impressed by how strongly you believe. It is impressed by whether your belief survives contact with the world *after* you have removed your ability to move the goalposts. When you can say, calmly, "Here is what I think will happen, here is why, here is how I measured it, and here is what would change my mind," you've done something rare in finance.

And rare is where edges live.

4.2 Data Pitfalls: Avoiding Survivorship and Selection Biases

A good backtest can feel like finding a secret passage in a familiar building. The numbers line up. The drawdowns are civilized. The Sharpe ratio has that neat, respectable haircut. Then you try to trade it live and it behaves like a completely different animal. Most people blame the market for *changing*. Sometimes it did. Often, the market didn't change at all. Your history was edited.

Edited histories are seductive because they don't look edited. They look like "data." Columns, timestamps, prices, returns—the whole thing arrives with the quiet authority of a spreadsheet. But the market's past is not a museum exhibit. It's a crime scene. Pieces are missing, labels got switched, witnesses disappeared, and someone you've never met decided which facts were worth preserving. If you don't interrogate the data, you're not backtesting a strategy. You're backtesting a story about the past that may never have been true.

Bias is the catch-all word, but it's more vivid to think of it as a *camera angle*. Your dataset is a camera pointed at reality. The question isn't "is it accurate?" The question is "what does it systematically fail to capture?" Survivorship bias, selection bias, and look-ahead bias are three ways your camera can be pointed slightly wrong—and "slightly wrong" is enough to turn mediocre ideas into apparent money machines.

The most famous lesson comes from a place you wouldn't expect to learn about finance: WWII aircraft repair. The Allied forces wanted to add armor to planes returning from missions. They mapped bullet holes on the aircraft that made it back. The wings and fuselage were peppered with damage. The engines, strangely, were relatively clean. The naive conclusion was obvious: armor the wings and fuselage, because that's where the planes get hit.

Abraham Wald, a statistician, looked at the same diagram and saw a trap. Of course the engines were "clean." Planes hit in the engines often didn't return. The bullet holes on the survivors were telling you where a plane *could* be hit and still fly home. The missing planes were telling you where hits were fatal. Armor the engines.

That mental move—looking at what *isn't* in the dataset—is the skill. Markets produce survivors too. Funds that didn't survive don't show up in "top manager" lists. Companies that went bankrupt don't end up in today's major indices. Strategies that blew up don't get marketed with glossy tear sheets. If your analysis starts from a pile of survivors, you will systematically underestimate risk and overestimate skill.

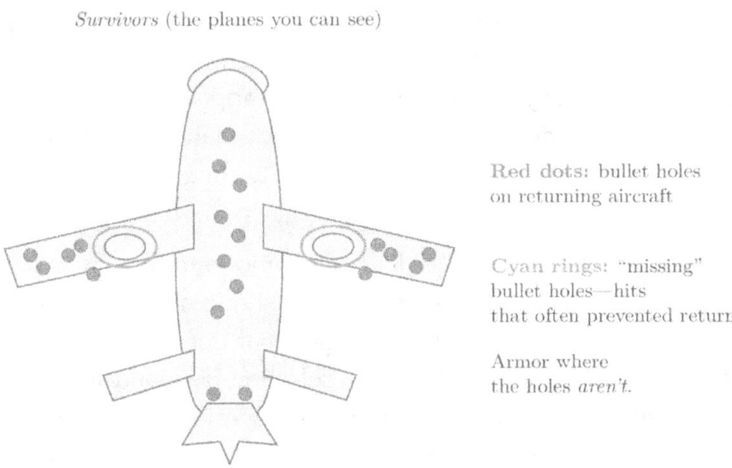

Survivorship bias in investing works the same way, except the wreckage is financial instead of metal. A classic example is mutual fund performance. If you evaluate "active managers" using a dataset that quietly drops dead funds, your sample becomes a hall of fame. Funds that performed badly often get merged away, liquidated, or renamed into something more marketable. If your dataset only contains funds that are still alive, you've accidentally asked the question: *How did the survivors do?* You did not ask: *How did the average investor's opportunity set do at the time?* That subtle difference can turn "active managers underperform" into "active managers look fine," or "persistence is weak" into "persistence looks real."

The danger is not academic. Survivorship bias changes the emotional feel of a strategy because it erases the ugly paths. A "low drawdown" backtest might simply have excluded the vehicles that experienced large drawdowns and disappeared. In other words, the dataset can function like a social media feed: all wins, no funerals.

Hedge fund data has its own special brand of optical illusion, partly because hedge funds are private, reporting is voluntary, and databases are stitched together from multiple vendors with messy incentives. One particularly nasty version is "backfill" or "instant history": a fund joins a database after a good run and reports prior returns, making its track record look longer and smoother than it would have looked in real time. Another is "hidden survivorship," where the mechanics of database construction effectively select funds with certain characteristics—often the ones that survived long enough to be captured cleanly. If you've ever wondered why hedge fund indices sometimes look like they discovered a loophole in capitalism, it's worth asking whether you're staring at a curated exhibit.

Selection bias is the bigger umbrella, and it shows up whenever the act of being observed depends on performance. Imagine a manager who only decides to launch a fund after a strong paper-trading period. Or a newsletter that only highlights "model portfolios" that worked. Or a factor research paper that explores a thousand variables and publishes the one that sparkles. The selection mechanism becomes a silent co-author of the results. You didn't just discover a pattern; you *selected* the pattern that survived your search process.

Here's a concrete way to feel selection bias in your bones: pretend you're evaluating traders by interviewing only those who are still in the business after ten years. You'll meet disciplined risk managers, sure. You'll also meet a handful of gamblers who happened not to hit the landmine. You won't meet the talented traders who took one oversized hit early and left the industry. You won't meet the mediocre ones who grinded until a regime changed and they didn't adapt. Your sample is contaminated by the simple fact that the losers are harder to find. In finance, being hard to find is not proof of being irrelevant; it's often proof of being important.

Then there's look-ahead bias, which is less like a missing plane and more like cheating with tomorrow's newspaper. It sneaks in when your backtest uses information that would not have been available at the time you're simulating. Some versions are obvious (using next quarter's earnings in today's signal). The scary versions are subtle enough to slip past smart people who swear they would never do that.

A classic subtle example is index constituent look-ahead. Suppose you backtest a strategy on "the S&P 500" from 1995 to 2025 using today's list of constituents. That is not the S&P 500 of 1995. It is a list of companies that, by definition, survived and thrived enough to be in the index later (or at least to be in your data vendor's reconstructed history). The companies that were kicked out, acquired at poor prices, or imploded might be missing or misrepresented. Your strategy is now trading in a past that has been sanitized by the future.

Even when you have the right constituent list, classification systems can smuggle the future into the past. Sector labels change. A company that becomes a "tech darling" might have been categorized differently years earlier. If you run a backtest that says "buy tech, short utilities" using end-of-sample classifications, you might be sorting companies based on what they *became*, not what they *were*. It sounds trivial. It isn't. Sorting by "what it became" is another way of selecting winners and calling it a signal.

Another look-ahead trap hides inside the word "fundamental." Fundamentals are reported with lags, restatements, and revisions. If you backtest using a clean dataset of annual financial statements without respecting the reporting date (and without accounting for restatements), you're effectively assuming you had perfect, final numbers at the time. Real investors don't. They have preliminary filings, surprises, delays, and accounting clean-ups that arrive months later. A strategy that appears to calmly harvest "quality" based on pristine data may, in live trading, be trying to drink from a firehose of delayed, corrected information.

All these biases share a single theme: they make the world look *safer* than it was. They reduce drawdowns, smooth equity curves, and make risk-adjusted returns look more stable. That's why they're so dangerous. Investors rarely reject a strategy because the backtest looks too good. They reject it because it looks too bad. Bias pushes in the flattering direction, so it gets welcomed like a friend.

The practical response isn't to become paranoid and refuse to test anything. It's to treat data like you treat counterparties: trust is earned, and you want terms in writing. When someone hands you a dataset, the first question is not "how many rows?" It's "what gets excluded, and why?" Ask about delistings. Ask about corporate actions handling. Ask whether the universe is survivorship-free. Ask whether index constituent histories are point-in-time. Ask whether fundamentals are point-in-time with reporting lags and restatement handling. These questions are the adult version of checking that the parachute is attached.

If you build your own datasets, you inherit the same responsibility. A common self-inflicted wound is filtering the universe using information that itself depends on survival. "Only trade stocks with five years of history" sounds like a liquidity or stability filter. It is also a survivorship filter. "Only include funds with three years of returns" is not a neutral requirement; it systematically removes early failures. Sometimes you *want* those filters for investability, but then you must admit what you did: you changed the question from "does the strategy work?" to "does it work in the subset that already survived long enough to qualify?" That's a narrower claim, and it should feel narrower when you size it.

Selection bias also appears in the most innocent sentence in finance: "I tested a few ideas." A "few" can be a hundred if you count every small tweak you tried and then forgot. Different lookback windows. Different thresholds. Different universes. Different rebalancing days. Different cost assumptions. Each attempt is a lottery ticket. If you only keep the winning ticket and throw away the rest, you'll sincerely believe you found something special. This isn't a moral failure.

It's an accounting failure. You did not record the true number of bets you placed in research.

One way to fight that is to keep a research log that records not only what worked, but what you tried. The point is not to punish yourself; it's to keep your sense of evidence calibrated. If you needed 200 tries to find one "amazing" backtest, the backtest isn't amazing. It's expected. In a noisy world, some patterns will look extraordinary purely by chance. The only way to know whether you're seeing signal is to remember how hard you searched.

There's also a more human selection bias: the strategies you're willing to test are shaped by what you already believe, what you've recently experienced, and what your social circle is excited about. When volatility has been low for a year, people invent strategies that sell volatility. When momentum has worked, people discover momentum "with better risk control." When a regime changes, the research fashions change. That doesn't make the ideas wrong. It means you should be skeptical of your own timing. If the only strategies you can imagine are the ones that would have worked recently, your imagination has become a trailing indicator.

A useful mental exercise is to intentionally design the "enemy backtest": a version of your strategy that should *not* work if your story is true. If your edge is about liquidity, test it in the most liquid names where the effect should weaken. If your edge is about small-cap neglect, test it in mega-caps where attention is constant. If the strategy works equally well everywhere, you haven't found a mechanism; you've found a generic exposure or a data artifact. Reality has texture. True edges often have fingerprints.

Another tool is to ask whether your signal could be "explained" by an untradeable assumption. For example, a backtest might assume you can always borrow stocks to short at a fixed cost. In the real world, hard-to-borrow names can become impossible to short exactly when your model wants to short them most. Or your backtest might assume you can transact at mid-price. Real execution happens at bid/ask, and the spread is a tax that rises in stress—precisely when turnover strategies are most vulnerable. If your model's best performance comes during volatile periods but your cost assumptions don't worsen in volatile periods, you may be harvesting an accounting mistake.

This is where the WWII bomber lesson becomes a habit rather than an anecdote. The empty space matters. What does your dataset *not* contain?

- Securities that delisted and disappeared.
- Funds that closed, merged, or stopped reporting.
- Real-time constituent lists, classifications, and rebalancing mechanics.
- Realistic borrowing constraints and dynamic transaction costs.
- The messy timing of information: reporting lags, revisions, restatements.

Any one of these can manufacture an edge. Together, they can produce a backtest that looks like a law of nature.

The punchline is not "never trust backtests." It's that the first job of research is not to maximize returns; it's to remove illusions. A strategy that survives on honest, ugly data is more valuable than a strategy that dazzles on a polished dataset. Polished datasets are great for presentations. Ugly datasets are great for not going broke.

If you want a single rule that prevents most of the worst mistakes, use this: *force your backtest to live in the same information world you would have inhabited at the time.* Point-in-time universes. Point-in-time fundamentals. Conservative, regime-sensitive costs. No "today's winners" masquerading as "yesterday's universe." When the past is allowed to be as messy as it really was, many edges disappear. That's not discouraging. That's the market doing you a favor before you pay tuition in real dollars.

4.3 Signal or Noise: Practical Statistics for Investors

A profitable streak is the most persuasive liar in finance. Ten good trades in a row can make a mediocre process feel like destiny. Ten bad trades can make a sound process feel broken. The market doesn't need to trick you; your brain will do it for free. The hard part isn't learning new math. The hard part is learning to look at your own results and ask, without drama: *Was that skill, or was that the coin flip landing heads again?*

The previous work—writing hypotheses you can actually falsify, and scrubbing away the most common data mirages—puts you in the right arena. Now comes the part where many investors quietly lose their nerve: interpreting evidence when the evidence is small, messy, and noisy. You don't get courtroom-grade proof in markets. You get clues. Statistics, used properly, is just a disciplined way of not overreacting to clues.

Start with a blunt truth: performance is a sample. Not *your* performance in some moral sense. A sample, like tasting soup with one spoonful. The spoonful can be too salty, too bland, or perfectly balanced—and still tell you almost nothing about the pot. A year of returns is a spoonful. A quarter is a sip. A month is the steam coming off the surface.

Sampling variability is the name for this problem: even if a strategy has a real edge, short samples can look terrible; even if a strategy has no edge, short samples can look brilliant. This is not philosophical. It's arithmetic. In a world where day-to-day returns bounce around, the average you observe over a short window is dominated by randomness. Investors often treat "three years" as a solid amount of evidence because it *feels* substantial. Three years is 36 monthly data points. In a market that can swing 4% in a week, 36 points is not a mountain of proof. It's a handful of dice throws.

That's why "manager beats benchmark three years in a row" is not automatically impressive. Imagine a manager with a 50/50 chance of beating the benchmark in any given year (no skill, just noise). The chance of beating it three years in a row is $(1/2)^3 = 12.5\%$. In a world with thousands of managers, 12.5% is not rare. It's a crowd. Someone is always on a streak. The streak is not a scandal; it's what randomness looks like at scale.

This is where investors get trapped by a subtle emotional mistake: they treat probability as if it should feel surprising. But probability doesn't care about your surprise threshold. If you have enough trials—enough funds, enough strategies, enough crypto influencers with charts—the unlikely becomes routine. The right question is not "is this streak exciting?" The right question is "given how many shots were taken, how many streaks like this should we expect *even if nobody has skill*?"

That same logic applies inside your own research process. If you tested fifty variations of a strategy and one version looks fantastic, you don't have "one fantastic strategy." You have fifty lottery tickets and one winner. The winner might still be skill. But the default assumption should be that it's noise until proven otherwise. Your brain will want to crown it as the chosen one. Statistics is the part of you that asks to see the birth certificate.

Investors reach for the Sharpe ratio because it's a simple summary: average excess return divided by volatility. It sounds like the ultimate score. The catch is that Sharpe is an *estimate*. You don't observe the true average return or the true volatility of a strategy; you infer them from a finite sample. And estimates bounce around. A Sharpe ratio of 1.0 might not be meaningfully different from 0.7 if you only have a short history. In fact, it might not even be meaningfully different from 0.0 depending on the noise level.

The danger is that we treat Sharpe ratios as if they were heights measured with a ruler. They're not. They're heights measured while standing on a trampoline in a windstorm. Two managers can differ by 0.3 in Sharpe just because one happened to catch a few lucky months. The world is full of pitch decks built on tiny Sharpe differences that are, statistically, a rounding error with a nice font.

A particularly common sin is naive annualization. People take a monthly Sharpe and multiply by $\sqrt{12}$ as if this were physics. That shortcut assumes the monthly returns are independent and identically distributed—no serial correlation, no smoothing, no regime shifts. Real strategies often violate those assumptions. Illiquid assets, smoothed marks, option-like payoffs, and trend-following exposures can all create serial correlation, which makes the simple scaling misleading. The score looks cleaner precisely when the assumptions are dirtier.

If you want a more reliable feel for what you actually know, replace "What's the Sharpe?" with "How uncertain is this Sharpe?" Think of a confidence interval as the range of Sharpe ratios that are consistent with the data you've seen. If that range is wide, your "1.2 Sharpe" might actually be anything from "pretty good" to "not real." Most investors don't compute the interval. You can still respect the concept: short samples imply wide uncertainty; wide uncertainty means small differences are not decisive.

This is why professional gamblers obsess over number of trials, not just win rate. A poker player who wins for three nights might be on a heater. A poker player who wins for three thousand sessions is almost certainly doing something right. Markets are not poker, but they share this: edge reveals itself slowly, and noise reveals itself quickly—by pretending to be edge.

A useful mental model here is the "lucky monkey." Give a monkey a keyboard and enough time, and it will type a Shakespeare sentence. Not because the monkey has talent, but because the number of random attempts is enormous. Now replace the monkey with the global asset management industry. Replace the keyboard with strategies, factors, and parameter settings. Replace Shakespeare with a backtest that looks like genius. You don't need conspiracy. You just need volume.

That volume problem is why multiple testing matters. Multiple testing is the statistical term for "you tried a lot of things." The more things you try, the more likely you are to find a false positive—a relationship that looks real in sample but is just noise. In finance, where everyone is testing ideas constantly, the burden of proof for a "new anomaly" should be higher than in a world where only one experiment was run. A t-statistic of about 2 (roughly "statistically significant" in many settings) can be far too lenient when you've effectively searched a giant space of possibilities. The market is a carnival of false positives; the prize is for the person who can resist them.

Practical investors don't need to memorize t-stat cutoffs to benefit from this. They need the instinct: *If the result appears after a lot of searching, demand stronger evidence.* Stronger can mean longer out-of-sample performance, replication across markets, a mechanism that predicts when it should weaken, or a strategy design with fewer knobs to overfit. The point is not to worship statistics. The point is to stop getting hypnotized by "significance" that was purchased by trying enough variations.

There's another reason "significant" can be a mirage: monetizability. A signal can be statistically convincing and still not be tradable in the way the model assumes. The backtest might assume you can trade at the close with no price impact. It might assume unlimited capacity. It might assume you can short anything at a stable borrow cost. It might quietly benefit from stale prices or infrequent marks. The result is a "factor mirage": the math says the effect exists, the P&L says it doesn't. Statistics can tell you whether a relationship appears in the data; it cannot, on its own, guarantee that you can harvest it after costs and constraints.

So what does "practical statistics" actually look like when you're sitting at your desk, looking at your own curve?

First, treat every metric as a distribution, not a point. Your CAGR is not *"12%."* It's "something like 12%, plus or minus a lot, depending on sample size and regime." Your max drawdown is not "-18%." It's "-18% in the slice of history you tested, but plausibly worse." The humility here is not optional. Metrics computed on short histories are loud opinions, not stable facts.

Second, separate *frequency* from *magnitude*. Investors love to ask, "What's the win rate?" because it feels concrete. But a strategy can win 70% of the time and still be terrible if the losses are occasionally catastrophic. Another can win 40% of the time and be excellent if the wins are large and the losses are capped. A high win rate is often a hallmark of picking up nickels in front of a steamroller: selling options, selling volatility, running carry trades with tail risk. Those strategies feel like skill right up until the steamroller arrives. Statistics should push you to inspect the left tail, not admire the average.

Third, ask whether the performance could plausibly be luck given the number of tries. This applies to managers and to yourself. If you run ten strategies and one is up 30% while the others are flat, that's not proof of genius. That's what you should *expect* to see sometimes in a noisy environment. You can't think clearly about evidence without thinking about the opportunity set.

A simple way to operationalize that is to write down the *research degrees of freedom* you exercised. How many universes did you consider? How many lookbacks? How many filters? How many entry/exit rules? If you can't remember, assume the number is higher than you want it to be. The point isn't guilt. It's calibration: the more freedom you used, the more skeptical you should be of the best result.

Fourth, be suspicious of smoothness. Smooth equity curves are comforting; they also deserve cross-examination. Some strategies are genuinely smooth because they diversify well and have stable exposures. Others are "smooth" because of pricing artifacts, infrequent marks, or hidden option-like risk that doesn't show up until it does.

Smoothing can be real or cosmetic. The statistician's instinct is to ask: if this were truly low risk, why is the market paying so much for it?

Fifth, beware "just a little better." In real investing, tiny advantages can matter *if* they are stable and scalable. But tiny advantages are also where noise is most effective at impersonation. A backtest that improves Sharpe from 0.85 to 0.95 by tweaking three parameters is not automatically progress. It might be overfitting. It might be randomness. It might be a fragile alignment with one regime. If you can't explain why the tweak should work economically, the statistics are warning you: you may have optimized for the past's quirks.

That brings us to an underrated form of statistical thinking: *predicted failures*. If you have a mechanism, you should be able to predict where the edge will weaken. Trend strategies often struggle in choppy, mean-reverting markets. Carry strategies often suffer when funding stress rises. Quality tilts can lag in speculative manias. The goal isn't to have an excuse when it underperforms. The goal is to demonstrate that you understand the shape of your own distribution. When a strategy fails in the way you predicted, that's not fun, but it is evidence of coherence. When it fails in a way you never imagined, that's information too—usually about hidden exposures.

A practical habit is to keep a "noise budget." Decide, in advance, how much underperformance you will tolerate before you conclude that something has changed. This is not about stop-losses on a position. It's about stop-losses on a belief. Without a noise budget, you will abandon good strategies at the worst time and stick with broken ones far too long. With a noise budget, you can say, "This drawdown is within the range I expected," or "This is outside the envelope; I need to re-underwrite the hypothesis." That is statistical discipline with skin in the game.

All of this may sound conservative, but it actually makes you bolder in the right way. When you understand noise, you stop getting bullied by it. You stop doubling down because of a hot hand. You stop capitulating because of a cold streak. You start treating results as data points rather than judgments of your identity.

The paradox is that the more you respect randomness, the more you can build real conviction. Conviction doesn't come from a beautiful three-year chart. It comes from a strategy that survives repeated attempts to explain it away: it holds up across time, across plausible cost assumptions, across market regimes, and across the uncomfortable possibility that the best-looking thing you found might simply be the monkey typing Shakespeare.

Statistics won't hand you certainty. Markets don't offer that product. What statistics can offer is something more valuable: the ability to look at performance—your own or someone else's—and not be easily impressed. In investing, not being easily impressed is a competitive advantage. It keeps you from paying for luck and calling it skill. It keeps you from selling skill because it temporarily looks like luck. And it keeps your decision-making anchored to evidence rather than adrenaline.

4.4 The Backtesting Trap: Validating Your Edge Safely

Backtesting is the closest thing finance has to a time machine. You take an idea, drop it into history, and watch it trade through booms, panics, bubbles, and boring Tuesdays. That power is exactly why backtesting is dangerous. A time machine is also the perfect tool for cheating: you can wander around the past, pick up tomorrow's information, and return to claim you were always right.

The most common backtest failure isn't a coding bug. It's a *story bug*. You start with a hypothesis that's crisp enough to test, you gather data that's not obviously rigged, and then you do something very human: you look at the results and begin negotiating with them. You tweak the lookback "just a bit." You add a filter that "makes sense." You move the entry time from close to next open because "execution." Each adjustment feels like craftsmanship. In aggregate, it's often hindsight optimization: turning the past into a custom-made suit and then bragging that you look great in it.

A backtest can be honest and still be misleading because markets are generous with coincidences. With enough parameters, you can fit the wrinkles in any historical period. The fit will look immaculate *in sample* (in the data you used to tune the strategy) and fall apart *out of sample* (in data the strategy hasn't seen). This is overfitting: learning the noise, not the signal.

The emotional trap is that overfitting feels like progress. The equity curve improves. The drawdowns shrink. The Sharpe ratio cleans up. The strategy starts to look "professional." The problem is that the improvement is often purchased by turning your model into a highly specific answer to a highly specific question: "What would have worked in *that* exact history?" The market rarely repeats that exact history.

Think of backtesting as training a sniffer dog. If you train it in one room and then you keep rearranging the room until the dog finds the scent perfectly, you haven't trained a good dog. You've trained a room. Live trading is the next building over.

The simplest defense is also the one that gets violated most casually: separate training from testing. In-sample data is where you build and tune. Out-of-sample data is where you evaluate. If you use the test set to keep improving the model, it stops being a test set. It becomes more training data with a fancy label. The backtest will still look persuasive, because the human brain is excellent at forgetting how much it "helped."

A good research process has a slightly paranoid stance: assume you are, unintentionally, trying to fool yourself. Then build guardrails that make it harder. The guardrails aren't there to slow you down; they're there to keep you from driving off a cliff while congratulating yourself on speed.

One guardrail is to drastically reduce the number of knobs. Every parameter is a lottery ticket. A strategy with two parameters is not automatically better than one with ten, but it is harder to accidentally optimize into nonsense. If the edge is real, it should be visible without heroic tuning. When a strategy needs the RSI threshold to be 47 instead of 50, or the lookback to be 63 days instead of 60, you're probably fitting the past's fingerprint rather than the market's structure.

Another guardrail is to pre-register your choices—not in a public database, but in your own notes. Before you run the backtest, write down: the universe, the exact signal definition, the rebalancing schedule, the cost model, and what results would convince you *and* what results would cause you to reject the idea. That last part matters. Without rejection criteria, you'll keep sanding the strategy until it shines. Sandpaper is not research.

Costs deserve their own skepticism. Transaction costs are not a footnote; they are a regime. Spreads widen in stress. Slippage grows when everyone is rushing through the same doorway. Borrow gets tight exactly when you most want to be short. If your backtest assumes constant, friendly costs, it is silently assuming that your strategy is most profitable in the same moments when the market is most generous to execute. That's usually backwards.

A conservative cost model is not pessimism; it is honesty. If a strategy survives costs that feel a bit painful on paper, you've learned something valuable. If a strategy only works under a cost model that feels like a fairy tale, you've learned something valuable too—just not what you hoped.

The next trap is selection by "best backtest." Even if each backtest is individually honest, the act of choosing the best one creates a statistical illusion. If you run 500 variants of a strategy, one of them will look spectacular by chance alone. The more you search, the more likely you are to select a lucky outlier. Researchers sometimes formalize this as the probability of backtest overfitting (PBO): the likelihood that the model you picked as "best" is primarily a product of noise rather than signal. You don't need to compute PBO to understand its message: the best-looking backtest is the most suspicious one, because it had the strongest selection pressure.

This is where many investors get the logic backwards. They treat a dazzling backtest as evidence of strength. A statistician treats it as evidence of *having tried enough things*. Dazzle increases the burden of proof.

A practical response is to keep a "strategy graveyard." Every time you try a variation, log it. Not just the winners. Log the losers too. When you later admire the champion strategy, you can see the bodies it stepped over. That record keeps your confidence calibrated and makes it harder to forget how much optimization you did.

Walk-forward testing is another useful discipline when you can't resist iterating (and most people can't). The idea is to simulate how a strategy would have been developed in real time: fit on an early window, test on the next window, roll forward, refit, test again. You're not trying to get the highest backtested return; you're trying to see whether the development process itself produces something that holds up as time moves. Walk-forward doesn't eliminate overfitting, but it makes it harder to hide.

Cross-validation methods are cousins of this idea: repeatedly split the data into training and testing slices, fit on some, evaluate on others, and see how stable the results are. The important word is *stability*. Real edges tend to be boringly stable relative to noise. Noise is flashy, fragile, and highly sensitive to small changes in sample.

Backtesting also fails when the simulated strategy cheats on information timing even without realizing it. You already know the obvious versions. The subtle version is benchmark and universe timing: using end-of-period index constituents, modern classifications, or "clean" fundamental data that has been revised. These issues can creep in even when you think you are only backtesting price-based signals. If the universe selection itself is point-in-time wrong, the whole test is contaminated. The strategy may appear to trade liquid winners because the losers were removed before the test even started.

A good habit is to build your backtest as if you were *forced* to trade it. That sounds dramatic, but it's clarifying. If you had to allocate real capital and explain the results to an unimpressed committee, what would you demand from the simulation?

- Point-in-time data: constituents, classifications, and fundamentals as they were known then.
- A cost model that worsens in stress, not one that assumes the market stays polite.
- Limits on leverage and turnover that reflect the strategy's true implementation.
- Position sizing rules that prevent one lucky period from dominating the equity curve.

This is less about impressing anyone and more about preventing self-deception from wearing a lab coat.

The "committee" mindset also helps with another trap: narrative patching. When a backtest disappoints, you invent exceptions. "It didn't work in 2008 because that was special." "It didn't work in 2020 because that was special." "It didn't work in 2022 because that was special." Special periods are real, but if every failure gets its own exemption, your strategy is turning into a collection of excuses. The honest question is: is the strategy robust across the major regimes you care about, or is it a one-regime specialist?

A good robustness check is not to torture the strategy with dozens of random tests; it's to stress the assumptions that matter. If your strategy relies on mean reversion, what happens when volatility doubles? If it relies on stable correlations, what happens when correlations spike? If it relies on liquidity, what happens when spreads widen? Robustness is not about perfection. It's about not dying when the world stops cooperating.

Another underrated danger is "parameter herding." Many people discover similar strategies because they use similar tools and read similar research. This creates crowding not only in trades but in backtests. If everyone is optimizing on the same decades of equity data, the same anomalies, the same constraints, they may be converging on strategies that fit that particular history well. When the future differs, the herd gets disappointed together. Diversification in research methods can be as important as diversification in positions.

Even if you do everything right, a backtest is still hypothetical. That word matters. Hypothetical performance has a special power to mislead because it has no scar tissue. It doesn't include the missed fills, the partial fills, the operational errors, the "we didn't trade that day because systems were down," or the human decisions that show up precisely when the strategy is under stress. That doesn't mean you can't use backtests. It means you should treat them like prototypes, not finished products.

One way to bridge prototype to reality is paper trading or small-scale live testing with strict rules. Not because the P&L will be meaningful quickly, but because you'll discover the non-obvious frictions: data delays, signal timing mismatches, execution constraints, corporate actions weirdness, and the psychological cost of holding the position when the strategy looks stupid. Many strategies fail not because the edge wasn't there, but because the implementation was assumed rather than tested.

Near the end of the research process, there's a moment when you have to choose: keep optimizing, or stop. If you keep optimizing, you might squeeze out more in-sample performance while quietly increasing fragility. If you stop, you accept an imperfect curve and bet that the imperfection is the price of honesty. The better choice is usually to stop earlier than your ego wants. The ego loves a perfect backtest. The market loves to punish perfection.

The cliff is what it looks like when you don't stop.

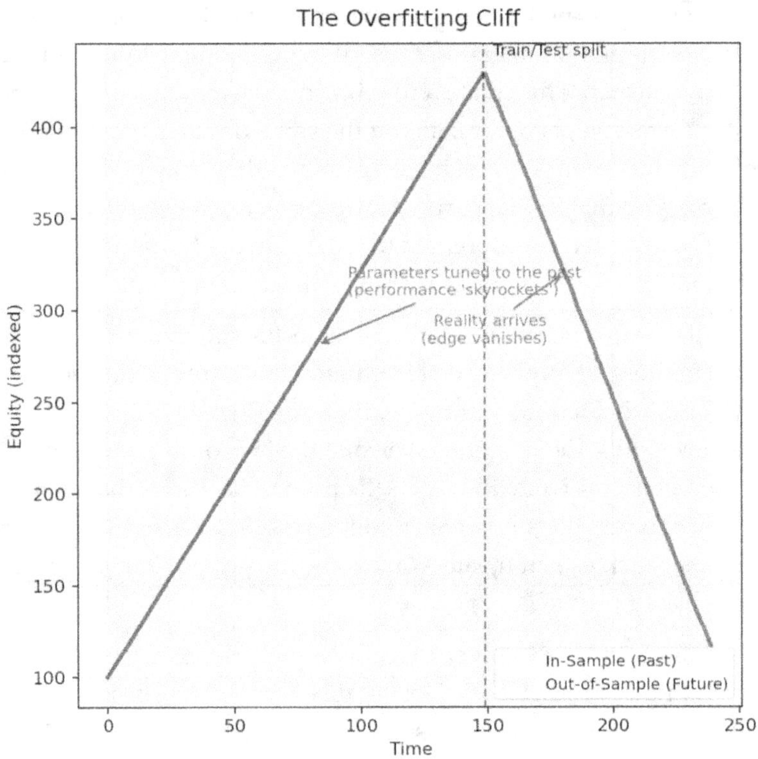

The cliff isn't just a quant problem. Discretionary investors fall off it too, just with different tools. They "optimize" narratives instead of parameters: they refine the story until the past becomes inevitable, then discover the future is under no obligation to follow the script. The antidote is the same: respect the boundary between what you used to learn and what you used to judge.

If you want a simple decision rule for when a backtest is mature enough to take seriously, try this: *Could I make it look this good if I were trying to deceive someone?* If the answer is yes—if the curve is too smooth, the drawdowns too polite, the improvements too dependent on precise parameter choices—then you should assume the strategy is fragile until it proves otherwise.

A safe backtest doesn't guarantee profit. Nothing does. What it buys you is something quieter but more valuable: you stop paying tuition for mistakes that were visible in the mirror. You trade fewer mirages. You size with more humility. And when performance disappoints, you can diagnose it with a process instead of a panic. That's what validation is really for: not to predict the future perfectly, but to keep you from lying about what you know.

Chapter 5

Managing the Bet: Sizing, Risk, and Execution

You've found an edge-a weighted coin in your favor. Now, the most dangerous question arises: how much do you bet? Bet too little, and you waste the opportunity; bet too much, and a simple run of bad luck wipes you out. This chapter shifts your focus from being a brilliant analyst to a disciplined risk manager. We will explore why position sizing determines your long-term wealth far more than stock picking, how to construct a portfolio that survives market earthquakes, and why the boring mechanics of execution are the final firewall between profit and regret.

5.1 The Goldilocks Zone: Sizing for Growth Without Ruin

A good bet can be turned into a bad investment by one decision: how much. That sounds almost insulting—as if the hardest part of investing is typing a number into the order ticket. But sizing is where the physics of compounding meet the psychology of pain. Get it right and you can be merely *ordinary* at forecasting and still build wealth. Get it wrong and even a real edge becomes a self-sabotage machine: too small and you waste time; too big and the market eventually finds the seam in your confidence and rips you open.

Most people treat position size like a personality test. Aggressive people bet big. Cautious people bet small. Professionals treat it more like engineering: given an estimated edge and a tolerance for turbulence, what fraction of capital produces the best long-run outcome *without* wandering into the region where one ugly sequence can end the game? That region exists. It's not drama. It's math.

Compounding is geometric, not fair

The first mental shift is to stop thinking in averages and start thinking in multiplication. If your portfolio goes up 50% and then down 50%, the average return is zero, but your money is not back to even:

$$1.5 \times 0.5 = 0.75.$$

You're down 25%. That's not a trivia question; it's a law of motion. Wealth compounds geometrically, which means volatility is not just uncomfortable—it's expensive. A sizing rule that looks "reasonable" in an arithmetic world can be quietly toxic in the geometric world where your actual life takes place.

This is why "expected return" can be a mirage if you ignore sizing. A strategy can have positive expected value and still be a terrible strategy *for you* if your bet size makes the path too violent to survive. Survival isn't a philosophical goal. It's a requirement for compounding to do its job.

The Kelly intuition: the size is part of the edge

The cleanest formal expression of geometric thinking is the Kelly criterion. The formula is not a commandment; it's a flashlight. It shows you where growth comes from and where ruin hides.

Start with a simple repeated bet. You risk a fraction f of your capital each time. With probability p you win, and your stake earns odds $b : 1$ (win b times the amount staked). With probability $q = 1 - p$ you lose your stake. After one bet, your wealth multiplier is:

$$\text{Wealth Multiplier} = \begin{cases} 1 + fb & \text{with probability } p, \\ 1 - f & \text{with probability } q. \end{cases}$$

Kelly says: choose f to maximize the *expected log growth*:

$$g(f) = p\ln(1 + fb) + q\ln(1 - f).$$

Differentiate, set to zero, and you get the classic fraction:

$$f^* = \frac{bp - q}{b} = \frac{(b + 1)p - 1}{b}.$$

That little star, f^*, is the *optimal Kelly fraction* under the model's assumptions: same bet repeated, stable probabilities, and no hidden tail risks waiting in the wings.

Even if you never plug in numbers, the intuition is gold:

- If your edge is tiny, your optimal bet is tiny.
- If the payoff is large relative to the loss (b is big), you can bet more for the same win rate.
- If your win probability is only barely above break-even, you're not "supposed" to bet big just because you *feel* confident.

The edge isn't just whether the bet is good. The edge is also whether you can size it so compounding can actually harvest it.

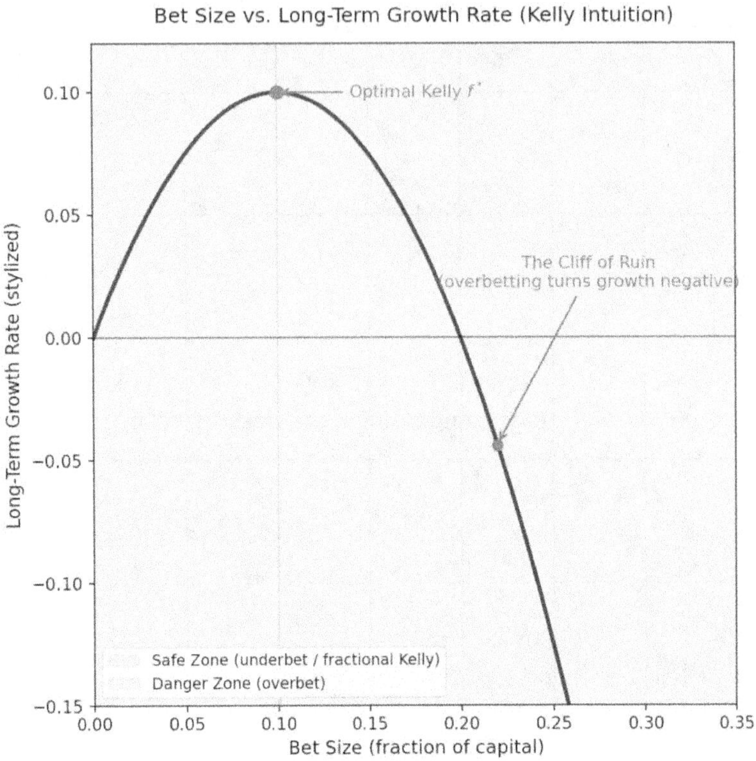

The shape matters more than the specific curve. Growth increases as you size up *until* it doesn't. There's a peak, and then there's a drop. Past the peak, you're not taking "more risk for more return." You're taking more risk for *less* long-run growth. And the penalty for overconfidence is not linear. It's a cliff.

Why overbetting is worse than underbetting

Underbetting is frustrating. Overbetting is lethal.

When you bet smaller than Kelly, you leave money on the table in exchange for smoother compounding. Your long-run growth rate declines, but it stays positive. You can be "wrong" about size and still survive.

When you bet larger than Kelly, something nastier happens: you can flip a positive edge into negative long-run growth. You can be right about the bet and still lose the game. Not in a single dramatic blow, necessarily. Often it's slower and more humiliating: a string of losses forces you to cut size, or triggers a margin call, or breaks your nerve so you sell the bottom. The market didn't defeat your idea; it defeated your *path dependency*.

A simple example makes the point. Suppose you take a 50/50 bet that doubles your stake if you win and loses your stake if you lose ($p = 0.5, b = 1$). This is a fair coin flip in expected value terms. Kelly says:

$$f^* = \frac{bp - q}{b} = \frac{0.5 - 0.5}{1} = 0.$$

The "optimal" long-run growth bet is to not play. Why? Because volatility with no edge is a compounding tax. You can't out-muscle the geometry.

Now give yourself a modest edge: say $p = 0.55$, $b = 1$. Then:

$$f^* = \frac{0.55 - 0.45}{1} = 0.10.$$

Ten percent. Not 50%. Not "all-in because I'm right." Ten percent. If that feels small, good: your intuition is colliding with the reality that a small edge cannot safely support a large fraction of capital. The market is a machine that turns small edges into wealth *only if* you don't demand instant gratification.

The hidden assumption: you actually know p and b

Kelly's derivation is clean. The world isn't.

The formula assumes you know the win probability and the payoff distribution. In markets, you estimate them through noisy data, changing regimes, and your own narrative bias. You think you have a 55% edge; maybe you have 51%. You think your downside is "limited"; then earnings gap down 28% through your stop. You think liquidity is "always there"; then it disappears exactly when you need it.

So the practical question is not "What is the Kelly size?" The practical question is: *How do you size when your edge estimate is wrong some meaningful fraction of the time?*

This is where the Kelly intuition stays valuable even when the Kelly fraction itself becomes untrustworthy. The shape of the curve doesn't go away. Overbetting still punishes you. Uncertainty about your edge is an argument for smaller f, not bigger. The more fragile your estimates, the more you should fear the right side of that curve.

Fractional Kelly: the professional compromise

If you ever wonder why so many serious investors sound boring about risk, this is why. They're not trying to maximize the return of a single trade. They're trying to maximize the return of their *career*, their *reputation*, and their *ability to keep playing*.

Fractional Kelly means betting a fixed fraction of the Kelly size, like half-Kelly or quarter-Kelly:

$$f = c \cdot f^*, \quad \text{where } 0 < c < 1.$$

The appeal is surprisingly asymmetric: you give up a little theoretical growth to buy a lot of practical survivability. The ride becomes smoother; the drawdowns become less savage; the odds that you panic at precisely the wrong moment drop.

Half-Kelly has a special charm because it tends to capture much of the growth while substantially reducing volatility of outcomes. Think of it as the adult version of confidence: "I believe in this edge, but I also believe I might be wrong about how big it is."

And there's a deeper benefit: fractional sizing is a way to honor *unknown unknowns*. Markets have tails. Models don't. When the tails show up, conservative sizing is the difference between a bad month and a permanent goodbye.

Sizing as an optimization problem, not a mood

Sizing feels like a tactical choice, but it's really your objective function made visible.

If your objective is to maximize *expected return* over a short horizon, you'll naturally drift toward big bets, because most edges look more exciting when you ignore the path. If your objective is to maximize *geometric growth* of wealth over many bets, you'll choose a size that respects the compounding penalty of volatility. If your objective is to maximize *sleep*, you'll choose a size that makes the worst plausible drawdown emotionally tolerable.

None of these objectives are "correct" in a vacuum. What's dangerous is pretending you have one objective while acting as if you have another. Many investors claim they are long-term compounders and then size like day traders chasing validation. The portfolio becomes a lie detector.

A useful practice is to write down, in plain language, what you're optimizing:

> "I am optimizing for the highest likelihood of compounding steadily over the next five years without experiencing a drawdown that forces me to abandon my process."

If that sentence is true, the right side of the growth curve should scare you.

A betting edge is not a permission slip to lever up

Here's a seductive thought: if you have an edge, leverage should be your friend. The math almost agrees—right up to the moment it doesn't.

Edges in markets are rarely stable. They decay when crowded. They invert when regimes change. They vanish when volatility spikes. Leverage is a multiplier on whatever is true; it doesn't discriminate between "true" and "true last quarter." The more your edge depends on a specific environment—low volatility, tight spreads, abundant liquidity, a particular correlation structure—the more leverage behaves like a trapdoor.

This is why "the bet is good" is not enough. You must also ask: *How wrong can I be, and still survive?* Full Kelly assumes you're not wrong. Fractional Kelly is what you do when you admit you're human.

The practical Kelly workflow: estimate, shrink, cap

There's a way to use Kelly thinking without turning your portfolio into a math contest.

Step 1: Translate the thesis into a distribution. Not a single target price. A range of outcomes with probabilities. If you can't do that, you don't have a bet yet; you have a story.

For a simple long trade, your rough ingredients are:

- Upside if right (how much can you plausibly make?)
- Downside if wrong (how much can you plausibly lose, including gaps and slippage)
- Probability of being right (and how confident you are in that estimate)

Even a crude distribution forces honesty. It makes you confront the difference between "I like it" and "I can size it."

Step 2: Compute an aggressive size, then assume you're overconfident. If you compute a Kelly-like fraction from your inputs, treat it as an upper bound. Then apply a haircut. Half-Kelly is the classic haircut. If your inputs are especially uncertain, quarter-Kelly is a sane default.

In practice, the haircut is doing more than reducing volatility. It's compensating for model risk: the risk that your p, your payoffs, or your independence assumptions are wrong. Markets love to attack assumptions.

Step 3: Add hard caps that reflect ruin dynamics. Kelly is about growth optimization, not operational survival. Your real constraints might include:

- Maximum single-position loss you can tolerate (financially and psychologically)
- Maximum portfolio drawdown before you must de-risk
- Liquidity constraints (can you exit without moving the market?)
- Correlated exposure across positions (ten "different" bets that all die in the same macro shock)

These caps are not admissions of weakness. They're acknowledgments that your portfolio is not a math problem floating in space. It lives in a world with margin rules, bid-ask spreads, and your own nervous system.

Sizing to the downside: the part everyone skips

Ask investors how they size, and you'll hear about conviction and upside. Ask traders how they size, and you'll hear about stops. Both groups often miss the same point: the downside is not just a number. It's a *process*. Losses change your future behavior.

A 10% portfolio drawdown might be annoying but tolerable. A 30% drawdown changes how you interpret information. Suddenly every headline feels personal. A 50% drawdown turns you into a different species: reactive, brittle, eager to "get back to even." The market doesn't need to bankrupt you to beat you. It just needs to push you into a mental state where you sabotage your own edge.

So a sane sizing method starts by asking: *What drawdown would cause me to violate my own rules?* Then you size so that drawdown is unlikely under your reasonable worst-case scenario.

That's not pessimism. That's professionalism. Optimism belongs in research. Caution belongs in sizing.

A memorable rule: bet big only when the edge is big

Kelly can be summarized in one sentence that's far more useful than the formula: *bet big only when the edge is big*. If the edge is modest, the correct size will feel modest. If you're tempted to bet big on a modest edge, that urge is not insight; it's impatience.

The market rewards patience in a peculiar way. It doesn't pay you for waiting. It pays you for *not forcing* the compounding engine to operate at an unsafe RPM. Fractional sizing is the governor that keeps the engine from blowing.

What ``ruin" looks like in real portfolios

In gambling textbooks, ruin is hitting zero. In markets, ruin has many disguises:

- A margin call that forces liquidation at the worst time
- A drawdown that triggers investor redemptions (if you manage money)
- A risk committee that cuts your limits right when your edge is greatest
- A psychological break where you abandon the process and start improvising

All of these are forms of terminal failure because they stop you from placing the next good bet at the right size. Compounding ends not only when the money is gone, but when the ability to execute the strategy is gone.

That's why the "cliff of ruin" on the growth curve is not melodrama. It's a picture of how overbetting converts temporary bad luck into permanent impairment.

The Goldilocks Zone

The Goldilocks Zone is not the maximum size you can □□□□ (tolerate) on a good day. It's the size that still works on a bad day. It's where:

- a normal losing streak doesn't force you to change your method,
- a volatility spike doesn't turn execution into a disaster,
- an estimation error in your edge doesn't push you off the right side of the curve.

If that sounds conservative, remember what you're buying: longevity. The market offers an endless series of bets. The scarce resource is not opportunity. The scarce resource is *staying solvent and sane long enough* for your edge to matter.

Once you see sizing this way, a lot of common behavior looks strange. "All-in" stops looking brave and starts looking like a confession that you don't expect to be around for many repetitions. Overconcentration stops looking like conviction and starts looking like a single point of failure. And a boring half-Kelly posture starts looking like what it really is: a quiet, disciplined commitment to letting time and compounding do the heavy lifting.

The best investors aren't the ones who are right the most. They're the ones who can keep placing good bets, at good sizes, through the full weather of randomness. That's the real edge: not just finding positive expectation, but engineering a path that can actually collect it.

5.2 The Only Free Lunch: Engineering True Diversification

Diversification has a branding problem. It's marketed like a multivitamin: take it daily, ask no questions, live longer. In practice, most "diversified" portfolios are a themed party where everyone shows up wearing the same outfit. Ten holdings, one bet. When the music stops, they all rush for the same exit at the same time.

The uncomfortable truth is that owning many tickers is not the same thing as owning many risks. A portfolio can look busy and still be fragile. The goal isn't to collect names; it's to engineer a set of return drivers that don't all bow to the same master.

A portfolio is a machine for mixing probabilities

Once sizing enters the picture, diversification stops being a moral virtue and becomes a mathematical tool. You're not just choosing what to own—you're choosing how the inevitable randomness in each position will combine. If the positions tend to win and lose together, you've basically built a single large position with extra administrative work. If they behave differently at different times, you've built a shock absorber.

That shock absorber is not a soft, feel-good benefit. It changes what you're *allowed* to do with sizing. A less volatile portfolio can often be sized more aggressively overall (or simply held with more patience), because it spends less time near the psychological and financial boundaries where bad decisions happen.

The difference between ``many" and ``different"

The phrase "own a basket" sounds prudent until you ask: a basket of what? Ten U.S. large-cap growth stocks is not a basket; it's a single factor bet expressed in ten dialects. The same is true of a portfolio stuffed with "defensive" consumer staples that all get hit together when rates jump, or with "value" stocks that all hinge on the same credit cycle. The holdings are different. The risks are not.

A simple way to think about it: every asset carries a few dominant sensitivities. Equity beta (how much it moves with the stock market). Interest-rate sensitivity (how it reacts to yields). Credit sensitivity. Commodity exposure. Liquidity exposure (how it behaves when everyone wants cash). If your portfolio loads heavily on one of these, you're concentrated—even if the tickers are alphabet soup.

True diversification is the act of deliberately *not* letting one storyline own your whole future.

Correlation: the hidden handshake between your positions

Correlation is just a number that describes how two return streams move together. But emotionally, it's the difference between "one thing went wrong" and "everything went wrong." When correlation is high, losses stack. When correlation is low, losses are more likely to arrive one at a time, which is a very different experience—and a very different compounding path.

The catch is that correlation is moody. It changes. It tends to rise when you most wish it wouldn't. In calm markets, many assets appear politely independent. In stress, they start behaving like a herd. That doesn't mean diversification is pointless. It means you should diversify with humility: assume correlations can spike, and build the portfolio so a correlation spike is painful but not catastrophic.

A useful mental model: diversification is not an on/off switch; it's a reduction in the probability that your worst-case scenario happens *all at once.*

The Markowitz punchline, in plain English

Harry Markowitz's big idea can be stated without equations: the risk of a portfolio depends on how the pieces fit together, not just how risky each piece is on its own. Two volatile assets can combine into something surprisingly stable if they tend to zig when the other zags. Conversely, two "safe" assets can create a nasty surprise if they share the same hidden dependency.

This is why looking at a single holding in isolation is like evaluating a brick without asking what building it's going into. A brick can be strong; a building can still collapse if the structure is wrong.

In portfolio terms, the "structure" is covariance: the way returns co-move. That's the plumbing. Ignore it and you get leaks.

Naïve diversification: the greatest hits album of hidden concentration

Naïve diversification is what happens when you diversify by category labels rather than behavior. It shows up in a few common costumes:

The sector sampler platter. A little tech, a little healthcare, a little financials. Looks balanced, until you realize the portfolio is still mostly a bet on U.S. equities doing well, with the same macro sensitivity running through everything.

The style mirage. Growth plus value plus quality plus momentum. Sounds like a well-rounded meal. Often it's just different wrappers around the same equity risk, and sometimes the factors that are "different" in theory become highly correlated in practice.

The international placebo. Add a handful of foreign stocks and declare victory. In global crises, many equity markets move together; you've diversified currency and politics, but not necessarily equity drawdowns.

The "alternatives" label. Private equity, venture, real estate funds, crypto, collectibles. Some of these can diversify, some can amplify risk, and many are linked by a single common trait: they depend on abundant liquidity and optimistic risk appetite. When those vanish, labels don't matter.

The lesson isn't cynical. It's precise: you don't diversify by counting. You diversify by understanding.

The only free lunch (with an asterisk)

Diversification is called the only free lunch because, in a world of uncertain forecasts, combining imperfectly correlated bets can improve the trade-off between return and risk. You can sometimes get a smoother ride without giving up much expected return. Sometimes you can even improve expected return if diversification allows you to stay invested through volatility instead of selling at the worst moment.

The asterisk is important: diversification is "free" only relative to the information you *don't* have. If you somehow knew with certainty which single asset would be the winner, diversification would look like self-sabotage. The free lunch exists because investors live in uncertainty and overconfidence is common. Diversification is a structural admission that your map of the future is drawn in pencil.

In other words, diversification isn't an argument against skill. It's an argument against pretending you have more skill than you do.

Correlation-aware sizing: stop treating positions as independent when they're roommates

The most practical upgrade investors can make is to stop sizing each position as if it lives alone. Two positions can each look "small," but together they can represent one oversized risk because they tend to move together.

Correlation-aware sizing starts with a blunt question: *If this goes wrong, what else goes wrong with it?* Not "what else is in the same sector," but what else shares the same underlying dependency. A semiconductor stock and a cloud software stock might look unrelated until you notice they both trade like long-duration assets sensitive to interest rates and risk appetite. A bank stock and a high-yield bond ETF might look different until credit spreads widen and they both get hit.

A portfolio that respects this reality often ends up with fewer "high-conviction" positions than your ego wants, and more true diversifiers than your attention span initially appreciates.

A simple way to think in portfolios: risk budgets

A risk budget is a decision to allocate not just dollars, but *volatility* (or drawdown risk) across exposures. Instead of saying "I'll put 10% in each of ten things," you ask, "How much of my total portfolio turbulence is each position allowed to contribute?"

This changes the conversation. A volatile, jumpy asset might deserve a small dollar weight if you want its diversifying behavior but don't want it steering the car. A stable asset might deserve a bigger weight if it quietly reduces overall variance and gives you room to take other bets.

Even without calculating portfolio variance formally, you can approximate risk budgeting with a common-sense heuristic: size inversely to typical volatility and adjust downward for high correlation with existing exposures.

The goal is not perfect optimization. The goal is to avoid the most expensive mistake: accidentally taking the same risk five times.

Diversification across regimes: build for different weather

Most portfolios are built for one season: the season investors recently experienced. That's why they break when the weather changes. True diversification isn't just owning assets that are different *on average*. It's owning assets that are useful in different regimes.

Think in terms of the conditions that dominate outcomes:

- Inflation surprise vs. disinflation surprise
- Rising rates vs. falling rates
- Growth acceleration vs. growth scare
- Liquidity abundance vs. liquidity stress

You don't need a crystal ball; you need coverage. A resilient portfolio tends to include exposures that can survive, and sometimes benefit, when the dominant narrative flips.

This is less about predicting the next regime and more about refusing to be held hostage by any single one.

Diversification isn't the same as diworsification

There is such a thing as owning too much. Add enough positions and you can end up with:

- higher costs and taxes,
- more complexity than you can monitor,
- a closet index that charges you active-level stress for passive-level returns.

That's not diversification; it's dilution. The point is not to eliminate all volatility. The point is to eliminate *uncompensated* volatility: risk you didn't mean to take, can't explain, and wouldn't choose if it were labeled honestly.

If you can't articulate what role a position plays—what it diversifies, what it hedges, what scenario it's meant to help with—it's probably not diversification. It's clutter.

The rebalancing paradox: the discipline that feels wrong

Diversification has an operational requirement: rebalancing. Without it, your portfolio slowly turns into a concentrated bet on what recently went up. That drift is sneaky because it feels like success. Your winners grow; your losers shrink; you tell yourself the portfolio is "naturally optimizing." Often it's naturally concentrating.

Rebalancing is emotionally backward. It asks you to trim what has treated you well and add to what has disappointed you. That's why it works as a discipline: it forces you to sell high and buy low in a systematic way rather than waiting for courage to appear.

Rebalancing also has a risk-control function that doesn't get enough credit. It keeps the portfolio's intended risk profile from quietly mutating. You may think you have a balanced machine, but without rebalancing you've built a machine that gradually shifts its center of gravity until it tips.

Diversification and the Kelly intuition: why smoother portfolios can grow faster

Here's the counterintuitive link: diversification can increase long-run growth by lowering volatility, even if it slightly lowers expected return. That's geometric compounding speaking again.

A smoother portfolio:

- suffers fewer deep drawdowns,
- spends less time in the danger zone where forced selling happens,
- allows you to size your *total* risk more confidently.

In Kelly language, diversification can move you leftward on the "cliff" curve without shrinking the opportunity set. It widens the Goldilocks Zone. It doesn't just make the ride more comfortable; it can make the destination more reachable.

This is also why "maximum Sharpe" (maximum return per unit of volatility) is more than a geeky metric. Lower volatility is not just lower drama. It can be higher geometric growth, because the compounding engine hates turbulence.

A practical test: the one-question diversification audit

If you only do one audit of your portfolio, do this:

> What is the single scenario that would hurt most of my positions at the same time?

If the answer is obvious—"rates spike," "credit spreads blow out," "a recession hits," "liquidity disappears," "the U.S. equity market sells off"—then you know what you're concentrated in. Concentration is not automatically wrong. It becomes wrong when it's accidental, unacknowledged, or sized as if it doesn't exist.

Then ask the follow-up question that turns awareness into engineering:

> If that scenario happens, what do I own that behaves differently?

Not "what do I own that I *hope* behaves differently," but what has a plausible economic reason to respond differently. If you can't name it, you've built a portfolio that's depending on one favorable climate.

Diversifiers should earn their seat

A true diversifier often looks unimpressive in isolation. It can have lower returns in bull markets. It can be boring. It can even be mildly annoying when your concentrated friends are bragging.

That's the point. A diversifier is not in the portfolio to win the popularity contest; it's there to keep the whole machine from vibrating itself apart. The right way to judge a diversifier is not "did it beat equities this year?" It's "did it do its job when the rest of the portfolio was under stress?"

This is why investors get diversification wrong: they evaluate components as if each must be a hero. In a resilient system, most components are not heroes. They are stabilizers.

The final trap: thinking diversification means you can stop thinking

Diversification is not a hall pass. Correlations change. Regimes shift. Hidden leverage emerges. Instruments that diversified yesterday can synchronize tomorrow, especially in crises. A portfolio is a living thing, not a static allocation.

But the right takeaway isn't paranoia. It's design. When you diversify intentionally—across drivers, across regimes, across liquidity profiles—you're not trying to predict the future. You're trying to avoid having your future determined by one mistake.

True diversification is a form of strategic humility with teeth: it accepts uncertainty, then builds a portfolio that can keep making good decisions inside it.

5.3 Defense First: The Math of Drawdowns and Survival

The market has a cruel way of grading you: not on how smart your ideas are, but on how much damage you can take without changing your behavior. Drawdowns—those stretches where your portfolio slides below its prior peak—are where good processes go to die. Not because the thesis was wrong, but because the investor became someone else.

The trap is that losses don't feel like math while they're happening. They feel like identity. They feel like betrayal. They feel urgent. That's why it helps to see the math cold, in daylight, before the storm hits.

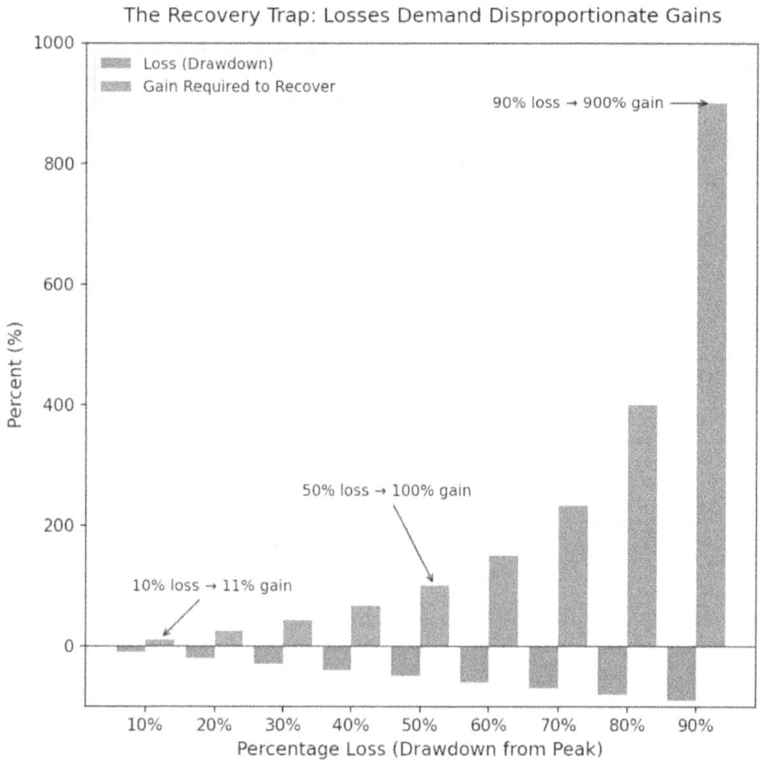

The chart isn't motivational; it's mechanical. Lose 10%, you need about 11% to get back. Annoying, but plausible. Lose 50%, you need 100%. That's not a "strong year," that's a doubling—and doubling is rare precisely when you're digging out of a hole. Lose 90%, you need 900%. At that point you're not investing; you're buying lottery tickets and calling it a plan.

This is why deep drawdowns are so dangerous. They don't just hurt. They *change the game you're playing*. The objective quietly shifts from "make good bets" to "get back to even," and that shift turns reasonable people into improvisers.

The asymmetry that never forgives you

The recovery math is simple. If your portfolio value drops from 1 to $1 - d$, the gain required to return to 1 is:

$$\text{Required Gain} = \frac{1}{1-d} - 1.$$

The ugliness is hidden in the denominator. As d approaches 1, the required gain explodes. That explosion is not a rare corner case; it's the natural consequence of allowing large losses.

This is also why "I can make it back" is often a delusion. The larger the loss, the more you must swing for the fences, and the more your process starts to resemble desperation. The market doesn't need to beat your expected value; it just needs to push you into a part of the payoff curve where you're forced to take bad bets to keep playing.

Survival is a portfolio constraint, not a personality trait

Some investors talk about drawdowns the way climbers talk about cold weather: proof of toughness. That's romantic, and wrong. Toughness is not a risk-management system. Drawdowns are not character-building in the moment they are building a margin clerk's confidence.

Survival is a constraint. It has financial dimensions (do you run out of capital or borrowing capacity?) and operational dimensions (do you violate rules, mandates, or client expectations?). It has a human dimension too: do you reach the point where you can't think clearly?

The market is full of strategies that "work" on paper but fail in real life because they assume an investor who doesn't flinch. That investor doesn't exist for long.

The two kinds of ruin: hard and soft

Hard ruin is the obvious one: you hit zero, or you get forcibly liquidated, or you can't meet margin requirements. It's dramatic and final.

Soft ruin is more common and more humiliating. It happens when you're still solvent but functionally out of the game:

- You cut risk at the bottom because you can't stand the pain.
- You abandon a strategy after a normal losing streak, right before it mean-reverts.
- You start "making it back" with oversized bets because patience feels like surrender.
- You stop trusting your own analysis, so every new piece of information whipsaws you.

Soft ruin is a behavioral margin call. Your account survives; your process doesn't.

The professional obsession with drawdown limits is not because pros hate losing. It's because they hate being forced into a different strategy by a bad month.

Drawdowns are not just outcomes; they are inputs

A subtle point changes how you size and build portfolios: drawdowns are not merely something that happens after you take risk. They should be a design input.

When you choose a position size, you're implicitly choosing a distribution of drawdowns. When you concentrate, you're choosing a distribution of drawdowns with fatter tails. When you lever, you're choosing a distribution with a shorter fuse.

That's why the "defense first" mindset isn't conservative in the weak sense. It's conservative in the way an aircraft engineer is conservative: the plane is meant to fly, but it must also land.

The simplest defense rule: cap any single blow

The easiest drawdown to avoid is the one that arrives in a single day. Single-day damage is uniquely poisonous because it doesn't give you time to respond thoughtfully. It forces action in a hot emotional state.

A practical defense is to set a hard cap on the maximum portfolio loss you can suffer from any *one* position under a plausible worst-case move. This is not the same as your stop-loss distance. Stops can fail. Gaps happen. Liquidity disappears. "Plausible worst-case" is the phrase that matters.

For an individual stock, plausible worst-case might mean an overnight earnings gap. For options, it might mean volatility expanding while price moves against you. For a crowded trade, it might mean a fast unwind where your exit is not your price.

If one position can plausibly take 15% out of your portfolio, you don't have a portfolio. You have a hostage situation.

Stop-losses: useful tool, dangerous religion

Stop-loss orders are often sold as an antidote to drawdowns: set the stop, go to sleep, wake up protected. Reality is messier.

A stop-loss is a pre-commitment to exit if price reaches a certain level. It can protect you from slow-motion denial, but it cannot guarantee your fill price. In fast markets, stops can trigger and execute materially worse than expected. In overnight gaps, the stop doesn't execute at your chosen level because your chosen level never traded.

This doesn't make stops bad. It makes them incomplete. The defense mindset treats stops as one layer, not the foundation. The foundation is position sizing that assumes stops may fail in exactly the moments you most want them to work.

The drawdown ladder: pre-decide your de-risking

One of the most expensive habits investors have is waiting to decide *how to respond* until after the portfolio is bleeding. That's like waiting to buy insurance until after the fire starts.

A drawdown ladder is a set of pre-decided actions tied to portfolio-level drawdowns. It's not about punishment; it's about preventing a bad stretch from escalating into a catastrophic stretch.

A simple ladder might look like:

- At −10% from peak: pause new risk, review exposures, check whether losses are thesis-driven or noise-driven.
- At −15%: reduce gross exposure (total risk) by a defined fraction; cut the least defensible positions first.
- At −20%: move to a capital-protection posture; tighten risk limits; trade smaller until the portfolio stabilizes.

The precise numbers are personal and strategy-dependent. The concept is universal: you need a rule that stops you from responding to losses with *more* risk. The human impulse is to press. The portfolio needs you to protect.

De-leveraging after losses: the compounding-friendly move that feels cowardly

After a drawdown, de-risking feels like admitting defeat. It also happens to be mathematically aligned with compounding. When your capital base shrinks, the same dollar risk becomes a larger percentage risk. If you keep the same position sizes, you are automatically levering up in percentage terms.

That's the silent killer. People think they're "holding steady," but in reality they are increasing their risk per unit of remaining capital.

The defense-first approach does the opposite: after losses, reduce risk so the portfolio's volatility stays roughly stable in percentage terms. This is not about timing the market. It's about preventing the drawdown from deepening into the recovery trap.

Stress testing: practice the pain before it arrives

Stress testing is the grown-up version of asking "what could go wrong?" It means running your portfolio through ugly scenarios and seeing if you still like your exposures.

You don't need a supercomputer. You need imagination anchored to history and structure:

- What happens if equities drop 30% quickly?
- What happens if rates jump and long-duration assets get repriced?
- What happens if credit spreads widen and liquidity thins?
- What happens if correlation spikes and your diversifiers fail at the same time?

The point isn't to predict which scenario will happen. The point is to avoid building a portfolio that can only survive one kind of world.

Stress tests also reveal hidden concentrations you didn't intend to own. A portfolio that looks diversified by ticker can reveal itself as a single macro bet when you shock the underlying variables.

The psychology of drawdowns: your future self is not you

When markets are calm, you can make beautifully rational promises. When you're down 18% and the headlines are screaming, your brain doesn't want optimization. It wants relief.

That's why defense has to be operational, not aspirational. It has to be encoded in rules, limits, and checklists. The purpose of those structures is not to restrict you; it's to protect you from the version of yourself who shows up under stress—the one who wants to do *something*, even if that something is a mistake.

Investors often overestimate their tolerance for drawdowns because they confuse *intellectual acceptance* with *emotional endurance*. The market tests the latter.

A calm definition of "risk limit"

A risk limit is a line you don't cross because crossing it turns a normal bad outcome into a potentially unrecoverable one. The best risk limits feel almost boring when you set them. They don't look heroic. They look a little too cautious. That's how you know they might work.

Good limits have three traits:

- They are stated in portfolio terms (how much can I lose?) not story terms (how much do I believe?).
- They are enforceable (you can actually execute them under stress).
- They are aligned with your true constraints: leverage, liquidity, time horizon, and psychological tolerance.

If a limit is unenforceable, it's not a limit. It's a wish.

The quiet power of staying near your peak

There's an advantage to avoiding deep drawdowns that rarely gets discussed because it sounds too mundane: you get to operate from a position of strength.

When you're near your equity peak, you're not chasing. You can be selective. You can wait for your edge. You can size thoughtfully. You can take losses without spiraling. You can hold a position through noise because noise doesn't threaten your identity.

When you're far below your peak, everything becomes harder. Your time horizon shortens. Your patience becomes expensive. Your risk-taking becomes reactive. The market can smell that desperation; it shows up in your fills, your timing, your decisions.

Defense is what keeps you in the part of the game where you can still think.

Defense first is not about fear; it's about optionality

Optionality is the ability to take advantage of future opportunities. It's what you lose when you suffer a crippling drawdown. A portfolio that survives has options: it can rebalance into bargains, increase exposure when the odds improve, and endure volatility without forced selling.

A portfolio that is broken has no options. It can only hope.

The paradox is that the more seriously you take defense, the more aggressive you can be when the bet is truly good. Not because you're reckless, but because you've protected the capital base and the decision-making apparatus that let you exploit the edge. That's what survival buys: the right to keep playing, with your mind intact and your wallet open.

And that is the real meaning of defense first. Not hiding from risk, but refusing to let risk end the conversation.

5.4 From Plan to Action: Execution and Discipline

The trade you *meant* to place and the trade you *actually* place are often different species. One lives in a calm, spreadsheet world where assumptions hold still. The other lives in a market that twitches, gaps, and occasionally panics like a crowd hearing a loud noise. Execution is the moment your research stops being a theory and becomes a set of irreversible clicks.

And execution has a particular kind of irony: you can do the hard intellectual work correctly, then donate the edge back to the market in the last thirty seconds through impatience, sloppy order choice, or a burst of ego that whispers, "Just get it done."

The market charges a spread for emotional urgency

Every investor has paid the urgency tax. You see the price moving, feel the window closing, and hit a market order like you're slapping a buzzer on a game show. You get filled, you exhale, and then you notice the fill price. The spread got you. The price moved. Maybe liquidity was thin. Maybe your order was larger than you told yourself it was.

That difference between the "decision price" in your head and the "fill price" in your account is slippage. Slippage is not a rounding error; it's the market's way of monetizing impatience. When your strategy's edge is measured in a few percentage points, consistently giving up a fraction of a percent on entries and exits is like carrying a leak in your boat and congratulating yourself on rowing harder.

The cleanest mental model is this: the market is happy to provide immediacy, but it is rarely free.

Cool-state planning versus hot-state clicking

When the market is closed—or when you're not staring at a flashing quote—you're a different person. You can think in distributions. You can respect sizing rules. You can accept uncertainty without trying to wrestle it to the ground.

When the market is open and your P&L is changing in real time, your brain starts doing something ancient and unhelpful: it treats price movement like immediate feedback on your worth. The same trade plan that looked sensible two hours ago can suddenly feel "too slow" or "too cautious," which is how thoughtful investors become compulsive traders.

The cure is not willpower. Willpower is a finite resource and markets are open a lot. The cure is pre-commitment: making the important decisions while you're in a cool state, then turning those decisions into orders and rules that are harder to violate in the hot state.

Pre-commitment is how you trade your own psychology the way you'd trade any other risk: by designing around it.

Execution is risk management wearing a different hat

Investors often file execution under "plumbing." Traders know better. Execution is risk management because it changes your realized payoff distribution.

A perfectly sized position can become oversized if you chase a fill and enter above your planned price. A well-defined exit can become a catastrophe if you assume liquidity will be there and it isn't. A diversified portfolio can become correlated in practice if you're forced to liquidate multiple positions into the same stressed tape.

In other words, execution is not separate from sizing and defense. It is where sizing and defense either become real or remain aspirational.

Market orders: the blunt instrument

A market order says, "Fill me now at whatever price is available." It prioritizes speed over price. That sounds fine when markets are deep and calm. It becomes dangerous when:

- the bid-ask spread widens,
- the order book is thin,
- the stock is moving quickly,
- you are trading around news,
- you are trading when liquidity is naturally worse (open, close, after-hours).

The risk isn't theoretical. The risk is that you give up far more than you intended, especially if you trade size. With market orders, you are outsourcing price discipline to whoever is quoting liquidity in that moment. Sometimes that's a rational counterparty. Sometimes it's an algorithm that smells desperation.

Market orders have a place. They are useful when getting filled matters more than the exact price—for example, exiting a position when your risk limit has been breached and the priority is to reduce exposure. But using market orders as your default entry tool is like buying a car by saying, "Just charge whatever." You will learn what regret feels like.

Limit orders: the scalpel (that sometimes doesn't cut)

A limit order says, "I'll buy at this price or better" (or sell at this price or better). It prioritizes price over immediacy. It's a way of insisting that the market meet *you*, rather than you chasing the market.

The obvious downside is non-execution. The price may run away without you. That can be painful if your thesis is right and you watch the trade work from the sidelines.

But that pain has information in it. If you miss a trade because you demanded a reasonable price, you didn't lose money. You lost participation. That's a different kind of loss, and it's often healthier. It prevents a common disease: turning "I want to own this" into "I must own this *now*."

Limit orders also force you to clarify your thesis. If you can't define a price at which you want to own something, you may be trading excitement, not edge.

Stop orders: the comforting story and the ugly reality

Stop orders are widely misunderstood. A stop order triggers when price reaches a stop level and typically becomes a market order. That last part matters. It means your stop is not a guaranteed exit price; it's a trigger for immediacy.

Stops can be useful for discipline. They can prevent you from negotiating with yourself while a position bleeds. They can act as a circuit breaker when your thesis is invalidated.

They can also behave badly in fast markets. During volatility spikes or dislocations, a stop can trigger and then fill far away from where you expected. In extreme conditions, stops can contribute to the very air pockets that create ugly prints. You wanted a safety net; you got a trapdoor.

The practical takeaway is not "never use stops." It's "treat stops as a tool with failure modes." If your entire risk plan depends on stops working perfectly, you don't have a risk plan. You have a bedtime story.

The execution frontier: paying in cost or paying in risk

Execution always forces a trade-off between two unpleasant things:

- *Market impact*: the cost you incur by moving the price against yourself as you trade (especially in size).
- *Price risk*: the risk that the market moves while you're trying to be patient and work the order.

This is the core idea behind the classic execution models used by institutions: there is an efficient frontier between cost and risk. Trade faster and you reduce the time you're exposed to adverse moves, but you pay more in impact. Trade slower and you reduce impact, but you risk the market running away or moving against you.

Retail investors often stumble into this frontier without knowing it exists. They either:

- smash market orders (minimize time risk, maximize cost), or

- set a heroic limit price and wait forever (minimize cost, maximize time risk and opportunity cost).

A disciplined executor chooses intentionally based on the situation: liquidity, volatility, size relative to average volume, and how sensitive the thesis is to entry price.

A simple execution rule: the more fragile the thesis, the more you need price discipline

Some trades are robust. If you're buying a diversified index allocation meant to be held for years, a few cents of slippage is not worth losing sleep over. Immediacy can be rational.

Other trades are fragile. If your edge depends on buying at a specific level, or if the position is highly volatile, or if you're trading around catalysts, sloppy entries can turn a good idea into a bad bet. In fragile trades, price is part of the thesis. You don't "kind of" execute. You either get your price or you don't get the trade.

The ability to walk away is a form of power in markets. Limit orders give you that power.

Pre-commitment in practice: write the trade as if you're leaving it to someone competent

A trade plan becomes real when it can be handed to a competent stranger and executed without further explanation. If the plan requires you to "feel it out" in the moment, it's not a plan; it's an invitation to improvise.

A robust plan specifies:

- the entry condition (not just the desire),
- the position size (not just "small" or "big"),
- the invalidation point (what would prove the thesis wrong),

- the exit logic (profit-taking, time-based exit, or thesis-based hold),
- the order type and execution approach.

Notice what's missing: hope. Hope is not an executable instruction.

Pre-commitment can be as simple as placing your limit order and your protective order (if you use one) immediately, then stepping away. The distance is the point. It creates a buffer between stimulus and response, which is where discipline lives.

Operational hygiene: the unglamorous edge

Most investors associate mistakes with bad analysis. In practice, some of the most painful losses come from operational errors:

- buying the wrong ticker,
- trading in the wrong account (taxable vs. retirement),
- forgetting an earnings date,
- misunderstanding the contract multiplier on options,
- accidentally doubling a position because you didn't notice a partial fill,
- placing an order on stale assumptions (news hit, volatility changed).

These aren't "market risks." They're self-inflicted wounds. The fix is boring: checklists, standard procedures, and a refusal to trade when you're distracted.

It's tempting to treat checklists as training wheels. They're closer to seatbelts. Serious participants use them because the cost of one preventable mistake can erase months of good work.

Best execution: your broker has obligations, but you have responsibility

In modern markets, your order may be routed across multiple venues. Brokers have a duty of best execution: to use reasonable diligence to seek the best market and achieve as favorable a price as possible under prevailing conditions. That regulatory obligation matters.

But it doesn't absolve you of the more important responsibility: choosing the right order type and the right level of urgency. A broker cannot rescue a poorly designed order. If you send a market order into a fast, thin market, "best execution" may still deliver a bad fill because the market you demanded is the market you got.

Think of best execution like a promise that the broker will try to treat your order fairly within the rules of the game. It is not a promise that the game will be gentle.

Partial fills, hidden liquidity, and why small investors still need respect for microstructure

Market microstructure is the study of how trades actually happen: order books, spreads, liquidity providers, routing. You don't need to become a specialist, but you do need to respect that prices are not magical numbers floating in space. They're the result of a matching process.

A few practical implications:

- A "last price" is not a promise. It's a report of the last match.
- A tight spread is a form of liquidity. A wide spread is a warning sign.
- A limit order can get partial fills, leaving you half-in and emotionally exposed.

- The visible order book is not the whole story; there can be hidden liquidity and dark pools, but you shouldn't rely on them for rescue.

When people say "the market is efficient," they rarely mean "the market is easy to trade." Execution friction is real. Treat it as part of the cost of doing business and you'll design around it instead of being surprised by it.

Discipline is mostly about removing choice at the wrong time

The romantic image of discipline is a hero resisting temptation. The realistic image is a person who designed their environment so temptation has less leverage.

If your process requires you to constantly make high-stakes decisions while watching the tape, you're turning investing into a stress test of your nervous system. That's not strategy; that's self-harm with a brokerage interface.

Better discipline looks like:

- fewer decisions made in real time,
- more decisions made in advance,
- more automation where appropriate (alerts, conditional orders, rebalancing rules),
- fewer opportunities to "just tweak it."

Tweaks are how plans die. They're usually justified with clever words and driven by adrenaline.

A final execution truth: the market is always offering you a deal; you don't have to take it

Markets are persuasive. They flash green and red numbers like they're reading your pulse. They offer you the illusion that action is safety. Often, action is just movement.

Execution discipline means you can watch the market offer you a worse and worse deal and still say, calmly, "No." You can insist on your price, your size, your risk limits, your conditions—and accept that sometimes the cost of doing it right is missing a trade.

Missing trades is not the enemy. The enemy is being so hungry for participation that you accept negative expected value in the form of bad fills, oversized entries, and exits driven by panic.

Good execution is quiet. It rarely produces a great story. It simply preserves the edge you already fought to find—and prevents the most common tragedy in investing: being right about the bet and wrong about the way you placed it.

Chapter 6

The Learning Loop: Measurement and Practical Playbooks

Profit is a fact, but skill is a theory. In the short run, a lucky gambler looks identical to a skilled trader; it's only over time that the truth emerges. This chapter moves beyond simple P&L statements to measure the actual quality of your decisions. We will build an honest scorecard that accounts for the 'silent partners' of fees and taxes, create a feedback loop that separates brilliance from dumb luck, and define clear playbooks for both the patient investor and the active trader to ensure you aren't playing a game you don't understand.

6.1 Keeping Score: Measuring Performance Net of Reality

A bull market hands out participation trophies. Prices rise, mistakes get upholstered, and even a mediocre portfolio can look like a personal masterpiece. The problem isn't that making money feels good; it's that the feeling is a terrible witness. If you want decision quality to improve over time, you need a scoreboard that refuses to flatter you. The cleanest name for that scoreboard is *True Alpha*: what your decisions added *after* you subtract what the market would have given you anyway, and after you pay the very real tolls of implementation—fees, spreads, slippage, and taxes.

Alpha is often treated like a brag, but it's really a residue. Think of your return as a glass after a party: what's left in the bottom after everything obvious has been poured out. The market's general upward drift? That's not you. Taking equity risk in a year when equities were strong? Not you either, at least not uniquely. Even "being in the right sector" can be nothing more than disguised beta if you loaded up on a factor the whole market rewarded. True Alpha is what remains after you account for (1) the exposures you intentionally or accidentally took, and (2) the frictions you paid to hold them.

Most investors keep score in a way that would get you fired in any other profession. A restaurant doesn't judge its chef by gross revenue; it judges by profit *after* food costs, labor, rent, and waste. Yet investors routinely celebrate a 14% year without noticing that the benchmark they could have owned in two clicks did 13%, their fund charged 1.25%, they churned the account into a short-term tax bracket, and they paid the spread every time they "just tweaked" the portfolio. A good scoreboard doesn't punish activity; it punishes *unpaid-for* activity.

The first step is deciding what game you are playing, because "beating the market" is not a single claim. If you own 80% stocks and 20% bonds, the S&P 500 is an unfaithful judge. If you own small caps, emerging markets, and a value tilt, the headline index is a costume party you didn't attend. Your benchmark should mirror the risk exposures you are willing to live with—not your aspirations, your self-image, or the index that happens to be on TV.

A practical way to do this is to build a *policy benchmark*: a simple mix of cheap, broad indices that matches your long-run target allocation. For a long-term investor, this might be 60/40 or 80/20. For a more diversified policy, it might be U.S. stocks, international stocks, and bonds in fixed percentages. The point is not sophistication; it's honesty. If your policy benchmark returned 10% and your portfolio returned 11%, you didn't necessarily produce 1% of skill, but you at least have a sensible starting line.

Once you have a benchmark, you can compute the most useful number in personal performance measurement: *excess return*. In plain language: how much you beat (or lagged) the benchmark over the same period.

$$\text{Excess Return} = R_{\text{portfolio}} - R_{\text{benchmark}}$$

This is where most "alpha" discussions stop, and where most self-deception begins. Excess return is *gross of reality*. It doesn't yet distinguish whether you outperformed because you took extra risk, whether you simply got lucky in a style that happened to work, or whether your outperformance evaporates once you pay the bill.

To strip away market tailwinds, you need to separate beta from alpha. Beta is the return you earned for wearing a known costume: equity risk, credit risk, duration risk, value, momentum, quality, size, and so on. If you don't measure beta, you will mislabel risk as skill. The classic way professionals do this is with a regression: you model your portfolio's returns as a combination of benchmark returns (or factors), plus a leftover term. The leftover term is the candidate for alpha.

The simplest version uses one factor—the market:

$$R_{p,t} - R_{f,t} = \alpha + \beta\,(R_{m,t} - R_{f,t}) + \epsilon_t$$

Here, $R_{p,t}$ is your portfolio return in period t, $R_{m,t}$ is the market return, and $R_{f,t}$ is the risk-free rate. The regression estimates β (your market sensitivity) and α (the intercept). The intercept is *risk-adjusted* excess return: what you earned beyond what your market exposure would predict.

But the one-factor model is an honest mirror only if your portfolio is basically "the market, plus or minus a bit." Many portfolios aren't. A dividend-heavy portfolio often carries value and quality exposure. A "growth stock" portfolio often carries momentum. A portfolio of small "disruptors" can be a cocktail of small-cap risk, unprofitable-growth risk, and a lot of narrative. In those cases, you can expand the model to multiple factors. You don't need to worship the exact academic taxonomy; the purpose is practical: don't congratulate yourself for simply being long what worked.

A helpful way to frame it: beta is the part of performance that a stranger could have replicated without knowing you. If a stranger could have matched your returns by holding a few broad ETFs that capture the same exposures, then your result is mostly beta. True Alpha is what a stranger *couldn't* get without your specific choices and timing, and what persists *after* costs.

Now the hard part: costs. Costs are where fake alpha goes to die.

Some costs are visible, like an advisory fee or a mutual fund's expense ratio. Others are quiet and embedded, like bid-ask spreads, market impact, and the price improvement (or lack of it) in execution. And then there are the costs that arrive wearing a friendly mask: taxes. Taxes don't feel like a fee because they're not labeled "fee," but they reduce your spendable wealth just as surely as a management charge.

Start with the easy subtraction: *explicit fees*. If you pay a 1% annual advisory fee and your portfolio returned 10%, your net return before taxes is closer to 9% (and in practice a bit less, because fees often apply to the account value regardless of performance, and because compounding works against you). If you own funds, their annual operating expenses are taken out of fund assets, which means you don't see a line item, but you still pay. The fee is there whether you notice it or not.

But explicit fees are only the tip of the spear. Trading has *implementation costs*: every time you buy, you usually pay the ask; every time you sell, you usually hit the bid. That gap is the bid-ask spread. Add commissions (if any), exchange fees, and the subtle but real effect of your own order moving the price when liquidity is thin. Even in "commission-free" environments, you can pay through spreads and execution quality. Your broker's marketing can be free; your fills are not guaranteed to be.

A useful mental model is to treat every trade as having a toll booth. Some tolls are pennies; some are dollars; some are a percentage point disguised as "a small slippage." If you don't track these, you will systematically overestimate skill in any strategy that trades frequently. Two investors can have the same paper strategy and radically different real-world results purely due to execution.

So how do you calculate a return that is net of these frictions? You build from cash flows. For an individual account, the cleanest measurement is: *how much money did I start with, how much did I end with, and what did I add or withdraw along the way?* The challenge is that contributions and withdrawals can distort the apparent return. If you add a large contribution just before a rally, a naive calculation will credit you for performance you didn't earn; if you withdraw before a rally, it can make you look worse than your choices deserved.

That's why professionals distinguish between *time-weighted* and *money-weighted* returns. Time-weighted return (TWR) is designed to measure the performance of the investments themselves, minimizing the effect of external cash flows. Money-weighted return (MWR), often expressed as an internal rate of return (IRR), measures the return on the actual dollars you put in, which is closer to your lived experience. Neither is "right" in all cases; they answer different questions.

If your goal is to evaluate *decision skill* as an allocator—what you bought and held—time-weighted is usually the fairer score because it doesn't give you extra credit for having more money invested at lucky times. If your goal is to evaluate your *real wealth outcome*, money-weighted is brutally honest, because it includes the timing of your own contributions, withdrawals, and risk exposure when it mattered to you. Many investors accidentally grade themselves with a messy hybrid that flatters their narrative.

You can use both, and you often should. The key is consistency: pick the metric that matches the question, and don't swap questions midstream when the answer is uncomfortable.

Now taxes—the most misunderstood performance drag because it's intermittent and emotionally "other." Taxes arrive in chunks, sometimes months after the decision, and they depend on holding periods, distributions, and the character of gains. The result is that pretax performance feels like a trophy, while after-tax performance feels like a penalty. But after-tax is the only performance you can actually spend.

A simple rule: your after-tax return is your pre-tax return minus the taxes triggered by your activity over that period, divided by the capital base. The devil is that taxes are triggered not only when you sell, but also when funds distribute dividends and capital gains, and when you harvest losses in a way that collides with wash sale rules. The tax code doesn't care about your intention; it cares about the transaction record.

Holding period is the hinge. In the U.S., gains on assets held one year or less are generally taxed as short-term capital gains, which are taxed at ordinary income rates; gains on assets held more than one year are generally taxed at long-term capital gains rates. That difference can turn a "nice trade" into a mediocre one after-tax, especially if you're operating in a high marginal bracket. Two strategies with the same pre-tax return can have wildly different after-tax outcomes simply because one churns and one compounds.

Taxes also create a strange form of performance mirage: a strategy can look brilliant in a tax-deferred account and merely decent in a taxable account. The strategy didn't change; the scoreboard did. That's not unfair. It's reality. If you care about True Alpha, you need to measure it in the habitat where it lives. The "same" portfolio can be a different animal after taxes.

A practical approach is to measure performance in layers, like peeling an onion without crying (too much). For a given period:

1. *Gross return*: what the portfolio did before any fees, trading costs, and taxes (often hard to know precisely for individuals).

2. *Net-of-fees return*: subtract advisory fees and fund expenses.
3. *Net-of-implementation return*: subtract estimated spreads, commissions, and slippage (harder, but approximable).
4. *After-tax return*: subtract taxes actually paid or accrued due to realized activity and distributions.
5. *Benchmark-adjusted after-tax return*: compare the after-tax result to an after-tax benchmark that reflects the same tax assumptions.

You don't need perfect precision on every layer to become dramatically more honest than the default investor. Even rough measurement is a spotlight.

One of the most useful tricks here is to compute what it would have taken for your active choices to be worth it. If your alternative is a low-cost index fund with a 0.03% expense ratio and minimal turnover, then every additional dollar of fee or friction you pay is a hurdle your decisions must clear *every year* just to tie. This is the "hurdle rate of activity." It's not moral judgment; it's math.

Put numbers on it. Suppose you run a $500,000 portfolio. You pay a 1% advisory fee ($5,000/year). Your funds average 0.40% in expense ratios ($2,000/year). Your trading costs (spreads, slippage, commissions) are a conservative 0.30% ($1,500/year). You've now created an annual drag of about 1.70% before considering taxes. To justify that drag, your choices must add at least 1.70% of gross outperformance—*and* do it persistently. If you're in a taxable account and you realize short-term gains, the hurdle rises again. Many investors are playing a game where their "edge" has to be professional-grade just to break even with a couch-potato index approach.

That realization is liberating if you let it be. It turns investing from an identity performance ("I'm the kind of person who picks winners") into a cost-benefit problem ("Is my activity purchasing enough expected value to justify its friction?"). The second question has answers you can live with.

Measuring implementation costs can sound daunting, but you can get surprisingly far with a few practical proxies. Track turnover: how much of the portfolio changed during the year. High turnover is a blinking sign that says "costs here." Track average holding period. Track how often you traded around earnings, open, close, or illiquid names. Track limit orders versus market orders. Most importantly, track your realized versus expected entry price: if your thesis assumed buying at $50 and you routinely paid $50.40 due to urgency and poor execution, that $0.40 is a hidden fee you chose.

Another underappreciated drag is *cash*. Idle cash feels safe, but it's a position with a return. If your benchmark is fully invested and you hold 10% cash for long stretches, your "underperformance" may not be poor security selection; it may be a timing bet on fear. Sometimes that bet is intentional and wise. Often it's just discomfort dressed as strategy. Either way, the scoreboard should force the admission: cash is not neutral.

Now for the most psychologically dangerous performance distortion: *selective measurement*. Investors love to measure what makes them look smart. They'll track the "realized gains" they booked and forget the opportunity cost of the winners they sold too early. They'll brag about a stock that doubled and ignore that it was a tiny position size that didn't move the portfolio. They'll anchor on the peak value of the account ("I was up 30% at one point") rather than the end value that pays the bills. They'll count dividends as "income" and ignore that the share price adjusted downward in the process. A rigorous scoreboard doesn't argue; it simply insists on the full record.

True Alpha, properly measured, is not a single number you check once a year like a blood pressure reading. It's a system: benchmark alignment, risk adjustment, cost subtraction, and after-tax realism. But you can still compress it into one practical calculation that fits in your head:

$$\text{True Alpha} \approx (R_p - R_b) - \text{Fees} - \text{Implementation Costs} - \text{Tax Drag}$$

Where each term is measured over the same period and on comparable assumptions.

Notice what this formula does emotionally: it takes away your ability to win arguments with yourself by changing the subject. If you beat the benchmark but only before fees, you didn't beat the benchmark. If you outperformed pre-tax but underperformed after-tax, you didn't create spendable outperformance. If you beat the market but only by taking more market risk, you earned a higher beta result, not necessarily alpha. That's not cynicism. That's clean bookkeeping.

There's also a subtle humility baked into good scorekeeping: sample size. Even if you measure perfectly, a year of outperformance proves almost nothing. A coin can land heads ten times in a row. Skill reveals itself slowly, and only if the measurement isolates it. That's why you should treat early alpha readings as *provisional*, like a medical test that needs repeat confirmation. The goal is not to declare yourself skilled; it's to create a measurement loop that can eventually tell the truth.

To keep this from becoming an exercise in self-criticism, tie measurement to decision usefulness. Good scorekeeping answers actionable questions:

- Are you being paid for the risks you're taking?
- Are your costs low enough that your edge, if any, has room to breathe?
- Is your activity increasing expected value, or just increasing the sensation of control?
- Is the tax bill the unavoidable cost of good decisions, or the avoidable cost of impatience?

When the scoreboard can answer those, it stops being judgment and becomes engineering.

One more practical point: benchmarks should be measured *net of what you could actually own*. Comparing a portfolio burdened with costs to a hypothetical, frictionless index is unfair in the wrong direction—it makes you look worse than the real alternative. The true alternative is a real, investable low-cost fund with a real expense ratio, real dividend distributions, and real tax consequences in your account type. A fair benchmark is not a Platonic ideal; it's the cheap, boring thing you can buy today and hold.

If you want to get especially rigorous, you can create two benchmarks: a pre-tax benchmark return and an after-tax benchmark return, based on your marginal tax rates and the benchmark's distribution profile. This is where many investors discover an uncomfortable truth: their active strategy isn't just competing against the market; it's competing against the market *plus* the structural tax efficiency of low-turnover indexing. In taxable accounts, tax efficiency is a form of edge. Ignoring it is like ignoring that one casino table pays worse odds.

None of this means active decisions are pointless. It means the bar is real. And once you see the bar, you can make smarter choices about where to spend your effort. Maybe your True Alpha isn't in stock selection at all. Maybe it's in asset allocation discipline. Maybe it's in avoiding behavioral blowups. Maybe it's in thoughtful tax management: harvesting losses, respecting holding periods, placing tax-inefficient assets in tax-advantaged accounts when possible. Those don't sound glamorous, which is precisely why they're available to ordinary investors.

The final move is to make the scoreboard personal without making it sentimental. Your True Alpha should be computed relative to your chosen risk posture and constraints. If you have a job that already gives you equity-like risk (say, startup compensation), then your benchmark might rationally be more conservative. If you have a short horizon, your benchmark should reflect that. "Beating the S&P" is not a universal mandate; it's a choice of arena. The best investors are not those who win every arena. They're the ones who choose arenas where their skills and constraints have a chance.

When you keep score net of reality, something paradoxical happens: investing becomes calmer. The need to constantly "do something" fades, because you can see exactly what activity costs you. You stop mistaking favorable weather for navigation skill. And when you do take an active bet, you take it with the quiet confidence of someone who knows the hurdle rate, knows the friction, and has decided the expected payoff is worth paying for. That's what True Alpha really is: not a victory lap, but a net profit after you've paid the world for the privilege of being in the game.

6.2 The Feedback Loop: Journaling and Skill vs. Luck

Markets are the only scoreboard most people have ever played where the loudest applause can follow the dumbest decision. A stock doubles because a meme caught fire. A reckless short survives because a central bank blinked. A "safe" bond fund quietly bleeds because inflation woke up. If you train your brain on outcomes alone, you end up learning the wrong lessons with enormous confidence. That's not a personality flaw; it's a natural response to an environment that rewards narratives more reliably than it rewards skill.

The most expensive bias in investing is hindsight. After the fact, the world feels explainable. You can always find a headline, a chart, a villain, a hero. The problem is that an explanation is not the same thing as a *forecast*. Hindsight doesn't merely edit your memory; it edits your identity. Your winners become proof of talent. Your losers become exceptions, "unlucky timing," or "manipulation." It's a cozy story—and a terrible training program.

A decision journal is the simplest tool that turns investing from entertainment into practice. Not a diary of feelings (though feelings matter), and not a scrapbook of trades (though records matter). A decision journal is a time-stamped claim about reality, written *before* the market votes on you. It's a way of pinning your future self to the wall, gently but firmly, so the post-game show can't rewrite what you believed when the bet was placed.

The core idea is separating *process* from *outcome*. Process is the quality of reasoning and discipline you used given what you could reasonably know at the time. Outcome is what happened. They are related, but they are not married. Great process can lose money. Bad process can make money. If you don't keep them distinct, you will become superstitious: repeating behaviors that happened to work and abandoning behaviors that happened to fail.

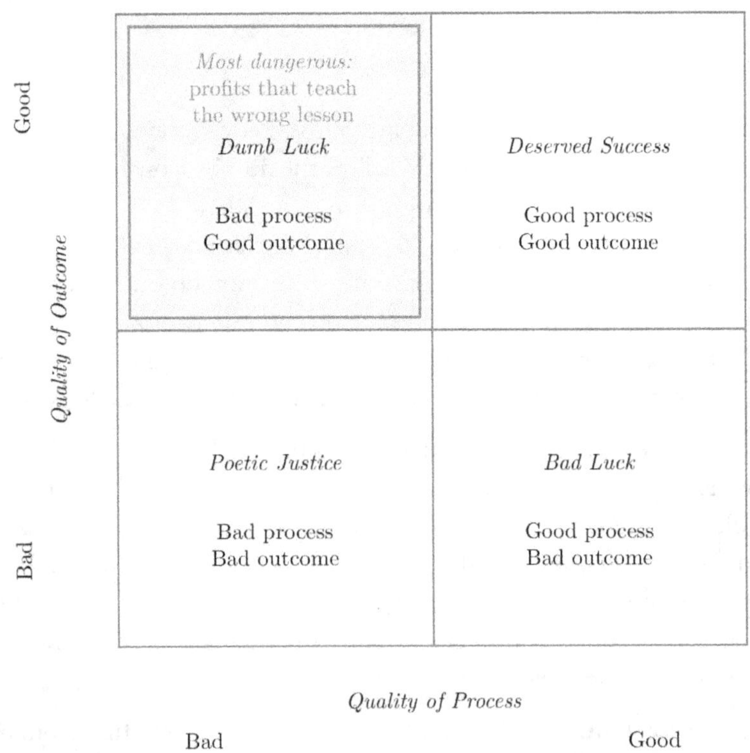

The "Dumb Luck" quadrant is where careers go to get quietly ruined. It feels like genius, which means you'll size up, loosen standards, and start believing you've discovered a repeatable edge. The market then does what markets do: it changes. When the tide turns, you don't just lose money; you lose your confidence and your map at the same time. A decision journal is partly a defense against losses, but more importantly it's a defense against *unearned* confidence.

Skill-versus-luck is a sampling problem dressed up as psychology. In a short window, randomness dominates. A handful of trades can make a terrible process look brilliant or a brilliant process look cursed. That's why professional performance studies often emphasize how rare persistent outperformance is, especially after costs. Most people intuitively grasp this when they hear it about mutual funds ("most managers underperform"), but forget it the moment they look at their own brokerage statement ("this *can't* be luck; it happened to me"). The journal brings the statistical humility home.

A good journal entry has two jobs. First: it captures what you believed and why, in language specific enough that you can later be proven wrong. Second: it makes evaluation possible by defining what "success" would look like *before* you see the result. That means writing down not only the thesis, but also the conditions under which you would admit defeat, reduce size, or change your mind.

The key is to journal the *decision*, not the security. "Bought XYZ" is not a decision; it's a transaction record. The decision is the rule you followed, the evidence you trusted, the alternatives you rejected, and the price you were willing to pay for uncertainty.

A practical decision journal template should feel a little like a pilot's checklist: not because you're trying to become robotic, but because you're trying to become reliable under stress. Here's a format that works for long-term investors and traders, with enough structure to be useful and enough freedom to capture what actually matters.

Field	*What to write (before you act)*
Decision snapshot	What are you doing (buy/sell/hold/resize), in what account, and why now? Include ticker, instrument type, and intended holding period.
Thesis in one sentence	A crisp claim about reality that could be wrong. If it can't be wrong, it's marketing copy.
Base rate	What usually happens in similar situations? Name the reference class (e.g., "small caps after a big run," "earnings gap-ups," "rate cuts with sticky inflation").
Catalyst vs. drift	Are you betting on a discrete event, or on slow compounding? If there's no catalyst, say so.
Your edge	Why should *you* be paid here? Faster info? Better process? Unique constraints? If the honest answer is "I don't know," that's data too.
Price & valuation anchor	What price are you paying, and what makes it cheap/expensive? For traders, replace with setup quality and location (levels, volatility regime).
Disconfirming evidence	What would change your mind? List the facts that would invalidate the thesis, not the feelings that would annoy you.
Decision rule	The rule you are following (rebalance band, breakout trigger, stop policy, maximum position size, tax constraint).
Risk & sizing note	Why this position size? What's the maximum loss you can tolerate without forcing bad behavior later?
Forecasts (with probabilities)	A few concrete outcomes with assigned probabilities and time horizons (e.g., "60%: flat to +10% in 12 months; 25%: -20% drawdown; 15%: +30%+").
Emotion and context	Sleep, stress, urgency, FOMO, need-to-be-right, recent wins/losses. A sentence is enough, but don't skip it.

The "base rate" line is where the journal quietly becomes a weapon. It forces you to ask: am I making a bet that has worked for other people over long samples, or am I improvising because a chart looks exciting? Base rates are not destiny, but they are the default gravity of the world. Ignoring them is how smart people end up making very confident, very expensive one-off bets.

The "your edge" line is the one most investors try to dodge with cleverness. They'll write something like "strong conviction" or "high quality company" or "this time is different." None of those are edges; they're adjectives. An edge is a reason to believe the expected value is positive for *you*, net of fees, taxes, and execution. Sometimes the edge is structural and unglamorous: low costs, patience, diversification, and the ability to rebalance when others panic. That's still an edge. It just doesn't make for good cocktail conversation.

Probabilities matter because they turn the journal into a calibration tool rather than a scrapbook. If you write, "I think it's going up," you can always claim you were "basically right" later. If you write, "I assign a 70% chance it's up at least 10% within six months," you've made a measurable forecast. After twenty of those, you'll know whether your 70% really means 70% or whether it means "I feel like it."

This is where forecast scoring comes in. You don't need an advanced statistics degree to benefit from it; you need a willingness to be scored. The simplest scoring method is the Brier score, which punishes both overconfidence and underconfidence. For a single forecast with probability p assigned to an event that either happens ($o = 1$) or doesn't ($o = 0$), the Brier score is:

$$\text{Brier} = (p - o)^2$$

If you say 90% and it happens, your score is $(0.9 - 1)^2 = 0.01$ (good). If you say 90% and it doesn't happen, your score is $(0.9 - 0)^2 = 0.81$ (painful, and deserved). Over time, you average these scores. Lower is better.

The beauty of scoring is that it changes what "being right" feels like. It rewards honest uncertainty. If you genuinely don't know, writing 55% is not weakness; it's accuracy. If you always write 80% because you like certainty, the score will eventually expose you as someone who confuses confidence with competence.

Calibration is the next layer. Group your forecasts by confidence bands: everything you labeled 60–70%, 70–80%, 80–90%. Then check how often you were right in each band. If your 80–90% band only hits 65% of the time, your brain is running hot. If your 60–70% band hits 80%, you're timid. Both distort decision-making: the first leads to blowups, the second leads to missed compounding.

The journal also reveals a more embarrassing truth: many "theses" are not theses at all; they're vibes. Vibes love wiggle room. They avoid time frames. They avoid conditions for being wrong. They avoid saying what would have to happen in the world for the trade to fail. A decision journal is anti-vibe technology. It forces specificity, and specificity is where learning lives.

A common objection is that journaling will slow you down. That's exactly the point. Most investing mistakes are not made because you lacked intelligence; they're made because you acted at the speed of emotion. The journal inserts a speed bump. If a trade can't survive two minutes of writing, it probably didn't deserve two months of capital.

For traders, the journal can feel like a threat because it exposes impulse. For long-term investors, it can feel unnecessary because the plan is "just hold." But long-term investors still make decisions: they deviate from allocation, chase performance, abandon rebalancing, concentrate in a story stock, panic-sell a drawdown, or quietly add risk because the last few years were kind. Those are decisions, and they deserve a record.

The post-mortem is where the journal turns into skill. The rules are simple and surprisingly hard:

- Review when the outcome is knowable (not when the position is merely up or down today).
- Grade the process first, without reference to P&L.
- Use the original entry as the standard, not your current memory.

A blameless post-mortem is not a self-esteem exercise. "Blameless" means you're trying to find broken mechanisms, not assign moral guilt. If you treat every mistake as evidence that you are stupid, you will stop looking. If you treat every win as evidence that you are smart, you will stop improving. The journal is a lab notebook, not a confession booth.

A useful post-mortem format is to force two separate grades:

Process grade: Did you follow your decision rule? Was the evidence sound? Did you size appropriately? Did you respect your constraints (taxes, liquidity, time horizon)? Did you identify disconfirming evidence in advance, and did you react when it appeared?

Forecast grade: Were your probabilities remotely calibrated? Did the outcome land in a bucket you considered plausible, or did it come from the "I didn't think that could happen" zone?

Notice what is not on the list: "Did it make money?" Money matters, but money is not diagnostic in the short run. The market can pay you to be sloppy, and it can fine you for being disciplined. The journal is how you keep your incentives aligned with reality instead of aligned with recent dopamine.

One of the sharpest uses of journaling is spotting repeatable failure modes. They tend to be boringly consistent. People break their own rules in the same ways:

- They increase size after wins, when luck has disguised fragility.
- They "average down" without new information, calling it conviction.

- They move the exit criteria after entry, calling it patience.
- They reinterpret the thesis after the fact, calling it nuance.

A journal makes these patterns undeniable because it creates a paper trail of your own justifications. The goal isn't shame; it's leverage. Fix one repeatable mistake, and you don't just improve a single trade-you improve a whole category of trades.

Journaling also helps with a quieter trap: confusing a good company with a good investment. A company can be wonderful and overpriced. A stock can be mediocre and undervalued. A journal entry that separates "I admire it" from "I expect a favorable payoff distribution at this price" is a small act of emotional hygiene that pays compounding dividends.

The illusion of control is the journal's main opponent. Investors love to feel like they are steering, and markets are happy to let you pretend—right up until they remind you that you were mostly along for the ride. The journal punctures the illusion by forcing you to name what you can control (process, sizing, costs, discipline) and what you can't (macro shocks, sentiment waves, sudden correlations). When you write those boundaries down, you stop wasting energy trying to dominate randomness and start investing in the parts of the system you can actually improve.

A good journal also keeps you from learning the wrong lesson from "bad luck." When a well-reasoned decision loses money, the reflex is to punish the behavior: "Never do that again." That's how people abandon sound strategies at the worst possible time. The journal lets you say something more mature: "The bet was positive expected value, the outcome was negative, and that is allowed." That single sentence is emotional durability. It prevents you from turning a normal drawdown into a strategy collapse.

The mirror image is more dangerous: "deserved success" should still be interrogated. A win doesn't mean the thesis was right; it means the trade made money. Sometimes you make money for the wrong reason. Maybe the catalyst you predicted never happened, but liquidity lifted everything. Maybe your valuation work was wrong, but sentiment bailed you out. A journal lets you separate "I got paid" from "I was correct." One is a result. The other is learning.

Over time, decision journaling turns investing into a personal dataset. That dataset becomes a brutally effective teacher because it contains what your brain normally deletes: your uncertainty, your motives, your constraints, and your original reasoning. After fifty entries, you can start asking higher-quality questions. Which setups produce good process but bad outcomes (suggesting a sound edge with variance)? Which setups produce bad process but good outcomes (a warning sign)? Which conditions reliably trigger your worst decisions (fatigue, urgency, recent losses)?

The final benefit is almost unfair: journaling makes you harder to manipulate. Markets, media, and social feeds run on urgency. They want you to act now, to pick a side, to feel certain. A decision journal is a private refusal. It says: if this is a real opportunity, it will survive contact with written reasoning. If it's merely excitement, it will collapse under the weight of a few honest sentences. Either way, you get your mind back.

Done consistently, the journal doesn't make you omniscient. It makes you *measurable*. And measurable is the first step toward getting better at a game where the world is always trying to convince you that the last outcome was your fault—or your genius.

6.3 The Patient Bet: A Playbook for Long-Term Investors

The hardest trade in investing is the one where you do nothing and still call it a decision. No button press. No chart. No dopamine. Just a quiet refusal to let the market draft you into its mood swings—the part of you that insists *something* must be done because *something* is happening.

The market's favorite trick is to make patience feel like negligence. When prices are ripping higher, sitting still feels like stupidity. When prices are collapsing, sitting still feels like denial. In both cases, the emotional experience is the same: urgency. Urgency is not information. It's just the body's way of saying, "I don't like uncertainty." The patient bet is the practice of treating that sensation as weather, not a forecast.

A long-term investor's core bet isn't "this stock will go up." It's "this *portfolio* structure will serve my goals across many market regimes, and I will stick to it long enough for the math to work." Your returns will still be volatile. Your emotions will still misbehave. The difference is that your system won't be built to react. It will be built to endure.

The playbook begins with one unglamorous object: an Investment Policy Statement (IPS). The IPS is a contract between the calm version of you and the future version of you who will, at some point, be convinced the world is ending or beginning. The goal isn't to predict crises. The goal is to pre-commit to a set of rules that remains sane even when you are not.

A useful IPS reads less like an essay and more like operating instructions. It answers questions that feel almost insulting to ask in good times: What is this money for? When will I need it? What drawdown would cause me to panic-sell? What asset allocation can I hold through an ugly year without changing my personality? If the IPS doesn't confront those questions, it's not a policy; it's a mood board.

Start with the target allocation. Asset allocation is a fancy term for a simple idea: decide in advance how much of your money is going to play offense (stocks and other growth assets) and how much is going to play defense (high-quality bonds, cash-like instruments, inflation hedges, depending on your situation). The market will try to make this feel like a mere "mix." It's not. Allocation is your portfolio's temperament. It determines how you will behave when the screen turns red.

A 90/10 portfolio and a 60/40 portfolio are not just different risk profiles. They are different psychological contracts. One asks you to tolerate violent drawdowns without flinching. The other asks you to accept that you'll look boring in melt-up rallies. Every allocation is a trade: you swap a certain kind of regret for a different kind. The IPS is where you choose your regret in advance.

Most investors pick allocation by imitation. They copy what a friend holds, what an influencer shows, or what looked good in the last five years. That's like choosing a winter coat by checking last week's temperature. A better approach is to pick an allocation that you can hold through a bad decade, not just a bad quarter. Decades happen. They don't send invitations.

Risk tolerance is often discussed as if it were a personality trait—"I'm aggressive," "I'm conservative." In practice it's a function of constraints. If you have stable income, a long horizon, and no near-term spending needs, you can afford more volatility. If your job is cyclical, your horizon is short, or your portfolio is supposed to fund something non-negotiable, volatility is not an aesthetic choice; it's a liability. The "right" allocation is the one that keeps you from making catastrophic behavior decisions.

This is where measurement and behavior fuse. The most dangerous portfolio is the one whose expected return looks great on paper but whose volatility forces you to sell at the worst possible time. That portfolio doesn't just underperform; it trains you to distrust your own plan. The patient bet is engineered around a simple objective: create a portfolio you can live with, so you don't have to keep reinventing it under stress.

Once you choose allocation, you need rules for maintaining it. Markets don't respect your target weights. They drift. When stocks rise, your equity percentage swells. When stocks fall, it shrinks. That drift is not a minor bookkeeping issue; it's the mechanism by which you accidentally change your risk level without noticing. A portfolio that quietly slides from 60/40 to 75/25 is not "doing well." It's becoming a different portfolio.

Rebalancing is the antidote. It means selling some of what has become overweight and buying what has become underweight to restore your chosen allocation. In other words, rebalancing forces you to do the emotionally unnatural thing: trim winners when you're feeling smart and add to losers when you're feeling scared. The headlines will never tell you to rebalance. Headlines are paid in clicks, and "restore target weights" is not clickbait. But rebalancing is one of the few behaviors in investing that mechanically encourages buying low and selling high.

There are two common ways to rebalance: time-based and threshold-based. Time-based is simple: you rebalance on a schedule, such as quarterly or annually. Threshold-based uses drift bands: you rebalance only when an asset class moves far enough away from its target weight, such as 5 percentage points or 20% relative drift. Neither is holy. The right choice is the one you'll actually follow, because the whole point is discipline, not aesthetics.

Rebalancing has costs, and pretending otherwise is how people turn a good habit into a leaky one. Trading incurs spreads and sometimes commissions. In taxable accounts, rebalancing can realize capital gains and trigger taxes. That means rebalancing needs to be implemented with some tact. You can rebalance using new contributions (buying what's underweight) instead of selling winners. You can use tax-loss harvesting opportunities to offset gains. You can prefer rebalancing inside tax-advantaged accounts where sales don't create immediate tax bills. The patient bet is not passive; it's *careful*.

The single most underused source of "alpha" for long-term investors is not security selection. It's *tax management*. Not sexy, but neither is flossing, and both work because they're boring and consistent. Taxes are one of the few drags you can reduce without predicting the market, and over decades the compounding effect is savage.

Asset location is the cousin of asset allocation, and it matters when you have multiple account types (taxable brokerage, IRA/401(k), Roth). Asset location asks: which assets belong in which accounts to minimize tax drag? Tax-inefficient assets—like taxable bonds throwing off ordinary income, high-turnover strategies that generate short-term gains, or REITs with distributions that may be taxed less favorably—often belong in tax-advantaged accounts when possible. Tax-efficient assets—like broad equity index funds with low turnover—often belong in taxable accounts where their deferred gains can compound. You can run the same "allocation" with very different after-tax results depending on where you park each piece.

A patient investor also learns to respect tax lots, the individual "batches" of shares you bought at different times and prices. Lots are not trivia; they're levers. When you sell, which lot you sell determines whether you realize a short-term gain, a long-term gain, or a loss. In a volatile market, careful lot selection can reduce taxes without changing your market exposure much at all. That's free money in the only way free money exists: the money you don't send away.

Wash sale rules complicate tax-loss harvesting, and that's precisely why a written plan helps. A lot of investors discover wash sales the way people discover black ice: suddenly and with regret. If you sell at a loss and repurchase the same or "substantially identical" security too soon, the loss can be disallowed and effectively deferred. That doesn't mean tax-loss harvesting is bad; it means it's a technique, not a vibe. The patient bet treats it like any other technique: defined rules, careful execution, and a record you can audit.

Now for the behavioral discipline that makes all of this work: the ability to do nothing while the market is screaming. "Screaming" takes different forms. Sometimes it's panic—a fast crash where your portfolio feels like it's leaking value every second. Sometimes it's euphoria—a melt-up where you feel foolish for not being more concentrated in the hottest thing. Both scream in the same language: *You are falling behind.* The patient bet is a refusal to accept the market's framing of time.

The market wants you to live on the timeframe where it is most persuasive: days and weeks. That's where randomness looks like meaning and where social comparison becomes unbearable. Long-term investing works when you drag your attention back to the timeframe where fundamentals and compounding have a chance to assert themselves: years and decades. You can't force the market to behave, but you can choose the horizon you use to judge it.

One of the most practical habits is to lower the frequency with which you check performance. This is not denial; it's risk management. If you check your portfolio ten times a day, you are effectively running ten emotional stress tests a day, and stress tests are not free. They drain discipline. The patient bet treats attention as a scarce resource and uses it where it pays: on behavior you control (contributions, costs, rebalancing rules, tax decisions), not on price wiggles you can't.

Another discipline is to pre-write your crisis script. When the market is down 30% and your brain starts proposing "temporary" moves, what will you do? "I'll be rational" is not a plan. A plan sounds like: "If equities fall by more than X%, I will rebalance within Y days using contributions first; I will not reduce equity exposure below target; I will review my IPS on a fixed date, not in the middle of a drawdown." The power comes from specificity. Your future self will be stressed and imaginative. Your script needs to be simple enough that stress can't negotiate with it.

A long-term investor also needs a policy for media. Financial media is optimized for engagement, not for your net worth. It will always have a narrative: recession, soft landing, hard landing, pivot, no pivot, AI boom, AI bubble, crisis, recovery, crisis again. The market is a complex adaptive system; narratives are simplified cartoons pasted on top after the fact. Treat them as entertainment unless they change your actual plan parameters. Most won't.

The patient bet doesn't require you to ignore reality. It requires you to distinguish between signal and theater. Reality is: your savings rate, your time horizon, your spending needs, your tax situation, your cost structure, your diversification, and your rebalancing discipline. Theater is: most daily commentary, most hot takes, and most "this changes everything" claims that arrive weekly and disappear quietly.

Diversification is another place where long-term investors accidentally sabotage themselves by seeking emotional satisfaction instead of risk control. Real diversification is not "owning many tickers." It's owning assets that behave differently under stress. If everything you own depends on the same macro factor (say, growth expectations or cheap liquidity), then you're not diversified; you're concentrated in disguise. Diversification won't maximize bragging rights in the best years. It exists to keep you solvent—financially and psychologically—in the worst years.

A patient playbook also includes guardrails against concentration creep. Concentration creep happens when one holding does well and becomes a larger part of your portfolio, and you decide not to trim because selling feels like "betraying" the winner. Over time, the portfolio stops being a balanced plan and becomes a personality cult built around one position. The market loves this because it turns discipline into drama. Your IPS should include position limits: maximum percentages for single stocks, single sectors, or single themes. Limits are not pessimism. They're an admission that no one knows the future well enough to bet the farm without eventually meeting a goat.

Long-term investors should also keep two return numbers in their head, because they answer different questions and prevent different mistakes. Your time-weighted return measures how the portfolio performed independent of your cash flow timing; it's a cleaner read on the portfolio's structure and the strategy's implementation. Your money-weighted return measures the return on your actual dollars, capturing the effect of when you added or withdrew funds—which is often another name for behavior. When your money-weighted return lags your time-weighted return, it's a clue that your timing decisions (contributions, panic pauses, performance chasing) may be the real leak. That's not an accusation; it's an opportunity.

The patient bet is also an ethical bet, whether you label it that way or not. Every dollar you invest is a vote for an economic future. Some investors want their capital aligned with their values; others don't care; many care but don't want the complexity. The playbook approach is to be explicit. If values matter, write constraints into the IPS (avoid certain industries, prefer certain practices, engage via proxy voting). If values don't matter to you, be honest about that too. The trap is pretending you have no values while implicitly adopting the loudest ones around you.

There's a final behavioral advantage to long-term investing that deserves to be stated plainly: patience is a competitive edge because it is scarce. Most people cannot tolerate being out of sync with the crowd. They buy when others are euphoric because being left out feels like being wrong. They sell when others panic because being alone feels like being reckless. If you can hold a sensible allocation through both moods, you don't need secret information to outperform the average investor's behavior. You just need emotional stamina and a plan.

That's the quiet scandal of markets: "average investor returns" are often worse than "investment returns" because humans are terrible at timing their own fear and greed. The patient bet attacks this gap with systems rather than sermons. It assumes you will feel the urge to act, and it puts your best self in charge ahead of time.

One last practical play: build a *patience budget*. Decide in advance how much "active tinkering" you're allowed per year, and treat it like calories. You can spend it, but you can't pretend it doesn't count. A patience budget might mean: one allocation review annually, rebalancing only when bands are breached, and no strategy changes unless there is a documented change in goals, horizon, or constraints.

This doesn't make you rigid. It makes you deliberate. It keeps your investing life from becoming a series of improvisations triggered by headlines.

The patient bet is not the absence of action. It's the presence of restraint. It's the decision to let compounding do the heavy lifting while you focus on the few levers that actually move the needle: costs, taxes, allocation, rebalancing, and behavior. When the market screams, you don't need to scream back. You need to check the policy, run the rule, and keep your hands steady. That's not passive. That's professional.

6.4 The Active Bet: A Playbook for Traders

Trading is the part of finance where the market stops pretending to be polite. Long-term investing gives you time to be vaguely right. Trading demands that you be right *on schedule*, under a microscope, while paying a toll on every decision. The market doesn't grade you on your thesis. It grades you on your fills, your sizing, your exits, and your ability to stay solvent long enough for your edge to matter.

The first misunderstanding to kill is the romantic one: an "edge" is not a hunch you've grown attached to. It's not swagger. It's not the feeling you get when three indicators line up like planets. An edge is a repeatable advantage that survives contact with costs—spread, slippage, commissions, financing, platform fees, and, in taxable accounts, the tax character of your gains. If it can't clear that bar, it's not an edge; it's a hobby with occasional good weeks.

A useful definition is brutally simple: edge is positive expected value. Expected value is the weighted average of outcomes: each possible payoff multiplied by its probability, summed up. Traders often get hypnotized by win rate ("I'm right 60% of the time") and ignore payoff distribution ("my losses are three times my wins"). Expected value forces you to look at the whole shape of the bet. A strategy can win often and still lose money. Another can lose often and still print, because when it wins, it wins big. The market doesn't pay you for being right; it pays you for having a favorable payoff distribution.

That distribution has to be measured net of reality. It's easy to back-test a strategy on mid-prices and pretend fills are frictionless. It's easy to ignore partial fills, fast markets, and the way your own order changes the tape. But trading is implementation. In practice, your "alpha" is often consumed not by being wrong about direction, but by being casual about execution.

So, what does a real edge look like? It usually comes from one of a few places, none of which are mystical. You might have a speed advantage (rare for retail). You might have an informational advantage (harder than people think, and often temporary). You might have an analytical advantage (better models, better pattern recognition, better regime identification). Or you might have a behavioral advantage: you execute cleanly while others hesitate, overtrade, revenge trade, or violate their own rules. For most discretionary traders, the edge is a cocktail of behavioral discipline and a narrow set of repeatable setups.

The word "narrow" matters. If your strategy is "trade what looks interesting," you don't have an edge; you have a personality. Edges are usually specific: a particular volatility regime, a particular liquidity condition, a particular reaction pattern around known events, a particular mean-reversion behavior in a defined universe. The narrower the claim, the easier it is to measure. Measurement is oxygen. Without it, your edge suffocates inside storytelling.

Even when you find something real, it decays. Edge has a half-life. In physics, half-life is how long it takes for half of a substance to decay. In markets, it's how long it takes for half of your advantage to disappear because other participants notice, copy, and compete it away—or because the underlying market regime changes. Traders who survive don't just discover edges; they manage the decay.

Decay happens for reasons that feel unfair but aren't personal. A strategy that worked in a low-rate, high-liquidity environment can die when rates rise and volatility behaves differently. A microstructure edge can vanish when order routing changes or when the crowd piles in. A pattern around earnings can change when guidance practices shift. A "risk-on/risk-off" playbook can fail when correlations break. The market is not obligated to keep paying you for yesterday's insight.

The antidote is to treat edge like inventory. You monitor it. You assume it spoils. You rotate stock. That means maintaining a rolling evaluation of performance that's tied to the strategy's logic, not just to P&L. If your setup is mean reversion, you should be tracking how far price typically stretches before snapping back, how long the snap takes, and whether the snap is still happening at all. If your setup is breakout momentum, you should be tracking follow-through rates, false breakout frequency, and how volatility interacts with your stop placement. Your journal becomes less about feelings and more about diagnostics.

A trader's journal is not a literary exercise. It's a lab notebook. The best entries are boring in the way a pilot's log is boring: the goal is to make success repeatable and failure explainable. Your journal should capture, at minimum:

- the setup category,
- the market regime you believe you're in,
- your entry trigger,
- your stop logic,
- your target logic (even if it's discretionary),
- your position size rationale, and
- what would have made you *not* take the trade.

When you review later, you're not hunting for excuses. You're hunting for patterns.

If that sounds like a lot, it is. Trading is a lot. The market is charging tuition every day. The only question is whether you're learning.

Execution discipline is the place where most traders lose their edge even when their ideas are decent. Discipline sounds like a moral virtue, but in trading it's operational. It's the ability to behave the same way on trade number 200 as you did on trade number 2, even when the last three outcomes have messed with your head. The market loves to exploit inconsistency because inconsistency creates predictable errors: chasing, early exits, late entries, widened stops, doubled size. These aren't "mistakes." They're failure modes. They show up the same way every time, and they can be engineered against.

Start with the non-negotiable: risk management. In trading, risk is not "volatility." Risk is the chance you can't keep playing. The market's most savage punishment is not a loss; it's a loss that forces you to liquidate, shrink into irrelevance, or go emotionally numb. Risk of ruin is the only risk that ends the game, which is why position sizing matters more than entry precision.

A clean sizing framework begins with a maximum loss per trade that you can tolerate without behavioral spillover. Some traders think in terms of a fixed percentage of equity (e.g., 0.25% to 1% risk per trade). Others use volatility-based sizing: position size scales down when volatility is high so that the dollar risk stays controlled. The key is consistency and realism. If your "stop" is so wide you won't honor it, it isn't a stop. If your size is so large you'll start bargaining with the market, your edge will be hijacked by your nervous system.

Stops themselves are misunderstood. A stop is not a guarantee of loss limitation; it's an intent. In fast markets you can gap through stops. In illiquid names you can get ugly fills. That's not a reason to avoid stops; it's a reason to assume slippage and size accordingly. The trader who sizes as if stops are perfect will eventually meet the day when they aren't.

Now the quiet killer: costs. Traders love "gross" thinking because gross thinking feels like talent. Net thinking feels like accounting.

Unfortunately, your broker and the IRS live in net world. Your edge must clear the friction hurdle *every time*. That means tracking at least three cost buckets:

- explicit costs (commissions, platform/data fees, margin interest),
- implicit costs (bid-ask spread, slippage, market impact), and
- tax drag (short-term gains, wash sales, special tax rules for certain instruments).

Many "profitable" strategies are just strategies that haven't yet met a realistic accounting of these costs.

Slippage deserves special attention because it's the cost you pay for urgency. Market orders buy speed and sell price. Limit orders buy price and sell certainty. Neither is morally superior. But you should know which one you're buying in each trade. If your strategy requires immediate entry, then slippage is part of the strategy, not a rounding error. Track it like you track P&L. A trader with a small edge who ignores slippage is like a restaurant that celebrates busy nights while forgetting it's undercharging for food.

Execution quality is also shaped by the plumbing you rarely see: order routing, venue selection, and the broker's incentives. Regulators talk about "best execution" because execution quality changes outcomes. For a trader, the practical implication is not to become a microstructure scholar overnight, but to become suspicious of vague fills. Compare your fills to the quoted prices at the time of execution. Track the difference. If you can't measure your implementation, you can't know whether your edge is real or whether it's being donated to the market in the form of poor execution.

There are also hard constraints that shape what kind of trader you can be, regardless of your ideas. In the U.S., accounts that meet the definition of "pattern day trader" face specific requirements and limitations, including minimum equity requirements and restrictions on day-trading buying power. This isn't a motivational speech; it's a boundary condition. Many small accounts blow up not because the trader is foolish, but because the strategy implicitly relies on leverage and rapid turnover that the account structure cannot support safely. If your plan assumes you can "just add" on margin, your plan is really "hope the broker and volatility are kind." Hope is not a risk control.

The market also punishes strategy/account mismatches. Some edges require holding through noise and letting trades develop, which is hard with tight margin constraints. Some edges require frequent small gains, which can be eaten alive by spreads and taxes. Some edges require trading liquid instruments with predictable microstructure; if you apply them to thin names, your backtest becomes fiction. A serious trading playbook starts by choosing an arena where your costs are low enough and your constraints are compatible with your holding period.

Managing edge decay requires regime awareness. "Regime" is just a fancy word for the market's current personality: trending or mean-reverting, volatile or quiet, correlated or dispersed, liquid or fragile. Most strategies work in some regimes and fail in others. The goal isn't to predict regimes with perfect accuracy. The goal is to notice when the environment has changed enough that your strategy's assumptions are no longer safe.

A practical method is to define a small set of regime indicators tied to your strategy. If you trade breakouts, you care about trend strength, volatility expansion, and breadth. If you trade mean reversion, you care about volatility spikes, dispersion, and how quickly extremes revert. If you trade event-driven setups, you care about gap behavior and follow-through. When those indicators move, you don't need to panic. You need to reduce size, tighten criteria, or stand down. Standing down is a position. It is often the highest-alpha trade you can make.

Standing down is psychologically hard because it feels like lost opportunity. But many traders don't fail because they can't find trades;

they fail because they can't *stop* trading. They confuse activity with professionalism. A professional is someone who can do nothing on a day with no edge and feel no shame about it.

A good playbook formalizes that with entry filters. Filters are not there to make you feel smart; they're there to keep you out of your own way. Examples include:

- only trade your setup in liquid instruments,
- only trade during defined hours,
- only take trades when volatility is within a band,
- only trade when the market regime matches your setup's historical sweet spot, and
- only trade when your last trade wasn't a revenge impulse.

The exact filters matter less than the fact that you have them and honor them.

The other half of discipline is exit behavior. Exits are where ego hides. Traders love to brag about entries because entries feel like prediction. Exits feel like admission. But your exit logic is the real expression of your edge. A strategy with a positive expected value can still fail if you systematically cut winners short and let losers linger. That is not bad luck; it's a negative payoff distribution you built with your own hands.

A clean exit policy is compatible with your setup. If you're trading a breakout, you might trail stops to stay in the move while protecting against a reversal. If you're trading mean reversion, you might scale out into the snapback and accept that you'll never sell the exact top. If you're trading around events, you might define a time stop: if the reaction doesn't happen within a window, you exit because the thesis is stale. The important thing is coherence. Random exits turn a structured bet into noise.

Taxes are the final "silent partner" that traders underestimate because taxes are delayed. If you trade frequently in a taxable account, most of your gains may be short-term, which can materially reduce your after-tax expectancy. That means the same pre-tax edge can be worth trading in a tax-advantaged account but not worth trading in a taxable one. If you ignore this, you can end up in the absurd position of being a decent trader who produces mediocre after-tax wealth.

The discipline layer that ties everything together is review. Not review as self-flagellation, and not review as victory lap, but review as refinement. Your journal should allow you to sort trades by setup type and evaluate them separately. Most traders are not uniformly good or bad. They are good at a few things and mediocre at many. The playbook's goal is to concentrate capital and attention where you are demonstrably competent, and to starve the parts of your trading that are basically entertainment.

A powerful review habit is to keep a "best trades" file and a "worst trades" file, but to label them by *process*, not by P&L. Your best trade might be a loss you took cleanly because the setup failed exactly as your rules said it might. Your worst trade might be a win you took impulsively with oversized risk. The market will pay you for that worst trade just often enough to keep you doing it—right up until it doesn't. The process lens protects you from that trap.

The active bet is a craft, and crafts are built from repeatable motions. Define the edge precisely enough to measure. Assume it decays, and monitor the decay without denial. Pay obsessive attention to execution, because implementation is where small edges live or die. Size so that you can survive variance. And treat discipline as operations, not virtue: rules you can follow even when your emotions have an opinion.

Trading can be a legitimate form of strategic decision-making, but only if you treat it like a business. Businesses track costs. They track process. They stop doing what no longer works. They respect constraints. The market doesn't care whether you call yourself a trader. It cares whether your edge survives in the only place that counts: the net result after reality takes its cut.

www.ingramcontent.com/pod-product-compliance
Lightning Source LLC
LaVergne TN
LVHW022214200926
848481LV00006B/1961
9798896653073